GMO CHINA

Contemporary Asia in the World

CONTEMPORARY ASIA IN THE WORLD

David C. Kang and Victor D. Cha, Editors

This series aims to address a gap in the public-policy and scholarly discussion of Asia. It seeks to promote books and studies that are on the cutting edge of their disciplines or promote multidisciplinary or interdisciplinary research but are also accessible to a wider readership. The editors seek to showcase the best scholarly and public-policy arguments on Asia from any field, including politics, history, economics, and cultural studies.

GMO China

HOW GLOBAL DEBATES TRANSFORMED CHINA'S AGRICULTURAL BIOTECHNOLOGY POLICIES

Cong Cao

Columbia University Press

New York

Columbia University Press
Publishers Since 1893
New York Chichester, West Sussex
cup.columbia.edu

Library of Congress Cataloging-in-Publication Data
Names: Cao, Cong, 1959– author.
Title: GMO China : how global debates transformed China's agricultural
biotechnology policies / Cong Cao.
Description: New York : Columbia University Press, [2018] |
Series: Contemporary Asia in the world | Includes bibliographical references
and index.
Identifiers: LCCN 2018012329| ISBN 9780231171663 (cloth) |
ISBN 9780231171670 (pbk.) | ISBN 9780231541091 (e-book)
Subjects: LCSH: Agricultural biotechnology—Government policy—China. |
Transgenic organisms—Government policy—China.
Classification: LCC S494.5.B563 C35 2018 | DDC 338.1/60951—dc23
LC record available at https://lccn.loc.gov/2018012329

Cover design: Lisa Hamm
Cover image: © Tony Hertz/Alamy Stock Photo

Contents

Abbreviations

APHIS	U.S. Department of Agriculture's Animal Plant Health Inspection Service
ARPA	Agricultural Risk Protection Act
BRAIB	Biotechnology Regulatory Authority of India Bill
BSE	bovine spongiform encephalopathy, also known as mad cow disease
Bt	*Bacillus thuringiensis*
CAAS	Chinese Academy of Agricultural Sciences
CAE	Chinese Academy of Engineering
CAS	Chinese Academy of Sciences
CASS	Chinese Academy of Social Sciences
CBD	Convention on Biological Diversity
CCAP	Center for Chinese Agricultural Policy
CCP	Chinese Communist Party
CCPCC	Chinese Communist Party's Central Committee
CCTV	China Central Television Station
CDC	Centers for Disease Control and Prevention
CFDA	China Food and Drug Administration
CGIAR	Consultative Group on International Agricultural Research
CJD	Creutzfeldt-Jacob Disease
CPPCC	Chinese People's Political Consultative Conference

CRISPR	clustered regularly interspaced short palindromic repeats
DBT	India Department of Biotechnology
DNA	deoxyribonucleic acid
EC	European Commission
EFSA	European Food Safety Authority
EPA	U.S. Environmental Protection Agency
ESA	U.S. Endangered Species Act
FSSA	India Food Safety and Standard Authority
EU	European Union
FDA	U.S. Food and Drug Administration
FFDCA	U.S. Federal Food, Drug, and Cosmetics Act
FIFRA	U.S. Federal Insecticide, Fungicide, and Rodenticide Act
FSSA	India Food Safety and Standard Authority
GAQSIQ	China General Administration of Quality Supervision, Inspection and Quarantine
GEAC	India Genetic Engineering Appraisal Committee, and later Genetic Engineering Advisory Committee
GEO	gene-edited organism
GM	genetically modified
GMA	U.S. Grocery Manufacturers Association
GMMP	genetically modified microbial pesticide
GMO	genetically modified organism
IARC	International Agency for Research on Cancer under World Health Organization
IHGP	International Human Genome Project
IP	intellectual property
IPR	intellectual property right
ISAAA	International Service for the Acquisition of Agri-biotech Applications
JMC	Joint Ministerial Conference, a mechanism within China's State Council
LMO	live modified organism
MEP	Mega-Engineering Program under China Medium- and Long-Term Plan for the Development of Science and Technology (2006–2020)
MLP	Medium- and Long-Term Plan for the Development of Science and Technology (2006–2020)

MNC	multinational corporation
MOA	China/India Ministry of Agriculture
MOE	China Ministry of Education
MOEF	India Ministry of Environment and Forests
MOEP	China Ministry of Environmental Protection
MOF	China Ministry of Finance
MOFCOM	China Ministry of Commerce
MOIIT	China Ministry of Industry and Information Technology
MOST	China/India Ministry of Science and Technology
MSP	Mega-Science Program under China Medium- and Long-Term Plan for the Development of Science and Technology (2006–2020)
NAS	U.S. National Academy of Sciences
NBCAGMO	China National Biosafety Committee on Agricultural Genetically Modified Organisms
NDA	India National Democratic Alliance
NDRC	China National Development and Reform Commission
NEPA	U.S. National Environmental Policy Act
NGO	nongovernmental organization
NHFPC	China National Health and Family Planning Commission
NIH	U.S. National Institutes of Health
NOC	no objection certificate
NOS	nopaline synthase
NPC	China National People's Congress
NSFC	National Natural Science Foundation of China
OAGEBA	China Office of Agricultural Genetic Engineering Biosafety Administration
OECD	Organization for Economic Co-operation and Development
OSTP	U.S. Office of Science and Technology Policy
PIP	plant-incorporated protectant
PLA	China People's Liberation Army
PPA	U.S. Plant Protection Act
R&D	research and development
rDNA	recombinant deoxyribonucleic acid
RNA	ribonucleic acid
S&T	science and technology
SAF	China State Administration for Forestry

SAIC	China State Administration for Industry and Commerce
SEPA	China State Environmental Protection Administration
SSTC	China State Science and Technology Commission
UNEP	United Nations Environmental Program
USDA	U.S. Department of Agriculture
VCU	value for cultivation and use
WHO	World Health Organization
WTO	World Trade Organization

Acknowledgments

This book is based on my long-standing interest in and extensive research on the development of life sciences/biotechnology in China, in the course of which I have incurred debts to individuals and organizations that have assisted me in various capacities. First of all, I would like to thank the interviewees for the study. Some of those involved in China's policymaking on the research and commercialization of agricultural genetically modified organisms (GMOs) took time from their busy schedules to receive me, several more than once. Although most of them are kept anonymous, their perspectives have been helpful in shaping the book.

The project was conceived in 2002 when, as a research fellow at the East Asian Institute, National University of Singapore, I first noticed the reports of imports of genetically modified (GM) soybeans into China. With a small grant from the university (R348−000−005−112), I undertook preliminary desk research and then initial fieldwork in China in the next couple of years.

In 2010, a grant from the U.S. National Science Foundation (NSF; SES−1115319) made it possible for me to further my research on policymaking pertaining to GMOs in China. The Institutional Review Board at the State University of New York (SUNY) Buffalo, through the introduction of Kathy Barberis at the SUNY Research Foundation, kindly helped with the research ethics clearance to meet a key prerequisite for the grant. In early 2011, I moved from SUNY's Neil D. Levin Graduate Institute of International Relations and Commerce in the United States to the University of Nottingham in the United Kingdom. With NSF's permission and the help of Thomas Moebus and Kathleen Bardolf at the Levin Institute, I was able to retain the grant awarded to one institution on one continent and use it at another university on another continent. The first draft of the manuscript was written in the second half of 2014 at Nottingham; finalization of the book, starting from the summer of 2015, corresponded to another move, this time to Nottingham's satellite campus in Ningbo, China, on yet another continent. I have thus benefited from the environments and the interactions with and encouragement and support from scholars at several institutions.

In the summer of 2017, a small grant from the Swiss National Science Foundation (IZK0Z1_175248) brought me to Switzerland, where I not only gained a European perspective on GMOs through talking to scholars and professionals there, but also was able to finish most of the final draft. I am especially grateful to Andrea Degen and Jessie Zheng Zhang of EURelations, who facilitated the visit, and Ralph Weber, my host at the Institute for European Global Studies, University of Basel, and his colleague, Cornelia Knab.

I also am very appreciative of the careful reading and critical but constructive comments on the entire draft offered generously and in a timely manner by Richard P. Suttmeier and Lester Ross. Hepeng Jia, a former science journalist in China who once covered GMOs and has been pursuing his Ph.D. in science communication at Cornell University, also took time from his studies to read and comment on several draft chapters.

Several Chinese scholars who have helped the conduct of the research and writing in various capacities happen to have the same last name—Liu. Lijun Liu, then with the Chinese Academy of Agricultural Sciences, not only contributed his expertise on the patenting of GM rice to a paper, now a chapter of the book, but also provided some much needed information timely. Yuxian Liu of Tongji University helped with research leading to the chapter on the newspaper coverage of GMOs. Xuxia Liu of Huazhong Agricultural University helped arrange interviews at her university in the summer of 2012. Meanwhile, Yutao Sun also provided critical help while pursuing his own academic endeavor.

Jonathan R. Cole, my Ph.D. adviser at Columbia University and now a colleague and friend, as well as Richard P. Suttmeier, Denis Fred Simon, and Rich Appelbaum, longtime collaborators and friends, among others, deserve special mention. Over the years, they have always provided professional, emotional, and moral support and encouragement whenever needed.

At Columbia University Press, Anne Routon, then commission editor of Asian studies, saw the value of the book project when it was only a proposal. She and her successor, Caelyn Cobb, waited patiently for delivery of the manuscript. Anonymous reviewers invited by the press read and commented on the proposal and an early draft of the book, providing very critical but encouraging comments whose influence is reflected in the current version.

Last but not least, I would like to thank my wife, Xiaozuo, and my son, Yiyang, for their patience with my research and writing. Without their love, support, understanding, and encouragement, it would have been impossible for me to endure more than a decade for a project like this.

Chapter 7 was originally published as Lijun Liu and Cong Cao, "Who Owns the Intellectual Property Rights to Chinese Genetically Modified Rice? Evidence from Patent Portfolio Analysis," *Biotechnology Law Report* 33 (2014): 181–192.

Introduction

Aseething debate over genetically modified (GM; genetically engineered, genetically altered, bioengineered, or transgenic [*zhuanjiyin*]) crops and foods has been unfolding in China. On March 1, 2014, Cui Yongyuan, a former anchor of China Central Television (CCTV, China's predominant state television broadcaster) and then on the faculty of the Communication University of China, uploaded a documentary on his fact-finding mission on GM foods in the United States for free viewing on China's three main internet portals—Sina, Tencent, and Sohu. Using one million renminbi (RMB) ($163,000) of his own money,[1] Cui shot the sixty-eight-minute film over a period of ten days, December 8–18, 2013, in Los Angeles, San Diego, Chicago, Springfield (Illinois), Seattle, and Davis (California). He visited supermarkets and farmers' markets and interviewed some thirty government officials, academics, campaigners in nongovernmental organizations (NGOs), and consumers to learn about the production of GM crops and consumption of GM foods in the United States. The documentary, which received millions of hits in a matter of days, was criticized as purportedly misleading by supporters of genetically modified organisms (GMOs), most notably Fang Zhouzi, a biochemist turned science writer who had been embroiled in a dispute with Cui over the issue since September 2013.[2]

The release of the documentary coincided with annual sessions (*lianghui*) of the National People's Congress (NPC), China's legislature, and the Chinese

People's Political Consultative Conference (CPPCC), a political advisory body, both of which convene in early March. As so often happens, Cui was a CPPCC member who later submitted several anti-GMO proposals at the meeting. These proposals called for an investigation into illegal planting of Bt (*Bacillus thuringiensis*) rice, a GM variety that had received biosafety certificates but had not been approved for production.[3]

This episode marked a climax of more than a decade of debate on GMOs in China. Since the turn of the twenty-first century, the nation has been caught in a tug-of-war between polarized anti- and pro-GMO camps. On one side stand well-known personalities such as Cui Yongyuan, as well as the international environmentalist NGO Greenpeace, some academics, and consumers. The anti-GMO camp does not simply oppose putting GM foods on Chinese dining tables; some activists promote organic and ecological farming over agricultural biotechnology. In addition to concerns over whether GM foods might cause health problems that go undetected for years and whether GM crops could damage China's biodiversity and environment, the anti-GMO group has trafficked in a nationalistic rhetoric that the United States has a hidden agenda or is engaging in a covert operation to harm the Chinese by introducing transgenic technology and exporting GM crops to China. The stakes seem highly significant to the Chinese, especially the political leadership.

On the other side, the GMO research community has been unrelenting in its insistence that the irrational rhetoric and actions taken by the activists have lost China momentum in the research and commercialization of GM crops. China started its GM-crop research and development (R&D) program in the mid-1980s. By 2000, it had become the most advanced among developing countries and was ranked fourth globally measured by arable land devoted to GM crops, with Brazil and India not even appearing on the list of leading GM nations. Since then, however, China has been overtaken by both Brazil, in 2003, and India, in 2006, and has experienced an overall decrease in GM-crop-growing land.[4] Scientists use the evidence that China has to import an increasing amount of soybeans—more than 95 million metric tons in 2017,[5] mostly genetically modified, from the United States, Brazil, and Argentina—to argue that this situation could have been prevented, or at least ameliorated, had domestic research and commercialization efforts been encouraged. They also claim that various issues raised by GMO skeptics are not only logically flawed but also politically sensational and irresponsible. As an example, leading Chinese agricultural biotechnologists have been labeled "American agents" worthy of scorn simply because of their scientific and professional origin in and/or scientific and professional associations with universities, multinational corporations (MNCs), and other organizations in the United States, such as Washington University in St. Louis, the University of California at Davis, Cornell University, Monsanto, and the Rockefeller Foundation. They have been character-assassinated and attacked on traditional media and increasingly on the internet and social media, and even harassed and humiliated in public.

In fact, the world is deeply divided over any given country's choice regarding GMOs—promotional, permissive, precautionary, or preventive—along policy dimensions of intellectual property rights (IPRs), biosafety, trade, food safety and consumer choice, and public research investment.[6] Most countries promote public R&D investment in transgenic technology, but their attitudes toward other dimensions of GMOs are quite different. For example, despite resistance, the United States has been permissive toward safety and consumer rights regarding GM foods. While preventive in IPR and trade, India has adopted permissive or precautionary policy choices regarding food safety and biosafety.[7] And the European Union has been at least precautionary, if not completely preventive, on most policy aspects.[8]

In the past three decades, as the world's largest agricultural country, and one of the most important, China has seen dramatic changes in its policy toward GMOs. Having consistently promoted public investment in R&D of GM crops to take advantage of the global biotechnology revolution, China has long been a frontrunner in the development of GM crops. It planted virus-resistant tobacco as early as 1988 and approved its commercialization in 1992, thus becoming the first country in the world to commercialize a GM plant. At one time, China's ambitions were to genetically modify "the majority of rice, wheat, corn, cotton, soy, and canola" by 2010.[9] It has been largely permissive in trade, as evidenced by imports of GM crops such as soybeans and corn in large quantities. However, its policy in other dimensions has been inconsistent. In 1999, the country was promotional in food safety and permissive in biosafety, more flexible than the United States, although its policy toward GM-food labeling has been among the most restrictive in the world. The introduction of Bt cotton for commercial planting from 1997 onward has been well received by cotton farmers. Domestically developed Bt-cotton seeds now dominate China's farmland, reversing an earlier monopoly by Monsanto, one of the leading agricultural biotechnology MNCs, although this is partly attributed to the restriction placed on foreign companies in carrying out R&D on GMOs and breeding GM seeds in China. Since entering the twenty-first century, China also has turned promotional through improving its weak IPR regime regarding GM crops.

A milestone was reached in August 2009 when the Ministry of Agriculture (MOA) issued five-year biosafety certificates for commercialization to Huahui 1 and Bt Shanyou 63, two strains of Bt rice, and phytase maize (corn). As well as overcoming a major hurdle for their planting,[10] the move signified that China could soon become the first country in the world to genetically modify rice, a major staple food. Nonetheless, the GMO policy later again became precautionary, if not preventive, regarding biosafety and food safety, and to a lesser extent, trade, thus significantly slowing the pace to develop and commercialize more GM crops. It is particularly notable that the commercialization of Bt rice and phytase maize stalled as the biosafety certificates were allowed to expire because of the leadership's inaction amid rising public concerns and other factors. (However, these certificates were renewed in early 2015 for another five years.) As a whole, China has thus far approved seven GM crops involving twelve transgenic events for commercialization, with only Bt cotton and papaya commercially available.

This book represents an effort to systematically document and analyze how China's policy toward research and commercialization of GM crops has evolved over the last three or so decades. Instead of taking the constructivist approaches of science and technology studies or anthropology,[11] it will focus on the complexity of the process through which policy has been put forward, debated and deliberated, formulated, modified, and implemented, and especially on how China's agricultural biotechnology community has set the agenda on research and commercialization while encountering resistance because of concerns over food safety and the rights of consumers, biosafety, IPR protection, and food sovereignty. In particular, this book is about how intertwined and multifaceted factors have influenced the evolution of policy toward research and especially toward commercialization of GM crops and foods, which in turn has affected the trajectory of transgenic technology in China. Moreover, against the backdrop of globalization, the shift of China's GMO policy has closely followed global trends, so that external pressure has also influenced this already multifaceted domestic policy issue. Given the size of its population and market, China has become a "battleground" where domestic as well as international stakeholders have engaged in heated "battles" over GMOs. Therefore, an examination of China's policy in conjunction with changing global attitudes toward GMOs will help achieve a better understanding of how China has interacted with the world.

The comprehensive analysis presented in the book has been made possible through the utilization and triangulation of data collected from various sources, including documents issued by the Chinese government and other stakeholders, scientific literature and patents, public discourse, and media coverage. More important has been the fieldwork conducted over more than a decade, between 2002 and 2014, in China, including some thirty interviews with scientists and other academics, government officials, policy analysts, activists, and journalists. Identified from scientific literature, media reports, and other sources, these interviewees were either contacted directly or introduced by other interviewees and colleagues. In the course of my research, I was able to assess the representativeness of the interviewees and to check the information collected from different sources to produce an accurate and balanced presentation.

This book has been written with a diversified audience in mind. As the development of transgenic technology in China has been embedded in a larger political, economic, and social context, and China's GMO policy is part of the leadership's policy portfolio, the book tries to provide a broader perspective of analysis to meet the interests of scholars of China studies. Scholars of science and technology studies will find the Chinese case a welcome addition to the analysis of dynamics and complexity in the development of an emerging technology with global political, economic, and societal ramifications. This book will also engage with analysts of science and technology (S&T) policy and government officials in an informed and intelligent dialogue about China's S&T policy and S&T development strategies in general, and GMO and agricultural biotechnology policy in particular. Finally, executives of agricultural biotechnology MNCs interested in doing business in China also will find this book informative and helpful.

In writing the book, I have chosen to begin by narrating the evolution of policy regarding transgenic technology and GMOs, globally and in China, and to defer conceptualizing and theorizing until the end. Specifically, chapter 1 details the global development of agricultural transgenic technology and the controversies around GMOs. Chapter 2 discusses GMO policies adopted in the United States, Europe, and India to establish a background for analyzing how China has been influenced by global trends in formulating and evolving its GMO policy. Chapter 3 situates the development of transgenic technology in China in a mostly promotional policy environment and then examines the R&D and especially commercialization of GM crops in China. Attention will be given to government policy, the role of foreign technology transfers embodied in human resources, and the efforts of domestic scientists to advance the technology. Chapter 4 analyzes the formulation and evolution of China's biosafety regulatory regime pertaining to GMOs and the rationales behind the establishment of such a regime, including associated risk assessment, in the context of globalization and China's evolution into a regulatory state. The cases of Bt cotton and Bt rice will be thoroughly examined in chapters 3 and 4, respectively. Discussing the tension between the research community and the public, chapter 5 focuses on how Chinese agricultural biotechnologists have played the roles of both technology developers and policy advocates, as well as how the public's response to the issuance of biosafety certificates to Bt rice, the staple crop, has reshaped policy toward research and commercialization on GM crops and biosafety regulations. The trial of Golden Rice among Chinese school-age children and its implications will also be examined. Chapter 6 explores how Chinese newspapers' coverage might have influenced the evolution of policy on GM crops and foods. Chapter 7 turns to patenting activities related to GM crops. By analyzing the patent portfolio of China's Bt rice specifically, this chapter will reveal not only the ownership of IPRs of the Bt rice but also the patent-infringement litigation prospects if China commercializes and exports Bt rice. Finally, in addition to summarizing the findings of the book and speculating where China will go in the near and medium future in its pursuit of becoming a GM nation, chapter 8 theorizes about the policymaking of post-academic science and post-normal science, as well as China's S&T policy cultures as they are applied to GMOs. Together, I hope these chapters paint a realistic picture of how China's changing views on GMO technology reflect its evolving position on the world stage.

GMO CHINA

CHAPTER I

Transgenic Technology and GMO Controversies

I f the eighteenth century was the era of steam, the nineteenth century electricity, and the twentieth century information technology, then the twenty-first century is likely to be the era of biotechnology. With the life sciences having seen some of the most exciting developments since the 1950s, life sciences–based innovation—biotechnology—holds enormous promise of providing solutions to some of the most challenging problems in the world, from population and health, to environmental degradation and climate change, to energy and agriculture. Such solutions may come in the form of radical, improved, and personalized drugs and therapies, novel medical devices and diagnostic toolkits, sustainable biofuels, efficient techniques for providing clean water and fighting pollution, and genetically modified organisms (GMOs) for agriculture.

Agriculture always involves genetic modification of plants, animals, or other organisms. For centuries, through selection, crossbreeding, hybridization, and more recently the use of radiation or chemicals, human beings have been able to induce random mutations to produce crops or livestock with desirable traits. Consequently, major domesticated crops or animals no longer resemble their wild ancestors. However, such a process is complicated, mainly involving the transfer and manipulation of genes between the same or neighboring organisms, as in the case of pollen carrying genes from one strain of rice to another. Traditional genetic modification also is time-consuming in terms of the breeding process, uncertain in terms

[1]

of controlling the types of genes introduced, and inefficient in terms of a process that is not targeted. In a word, it is largely a process of trial and error. The modern biotechnology used to produce GMOs is distinct from these breeding methods. "Genetic modification"[1] (or similar terms), as used in this book, refers only to the introduction of new genes into crops or plants through recombinant deoxyribonucleic acid (rDNA) technology. This chapter starts with a review of the development of transgenic technology, followed by a brief description of controversies around GMOs to set a technical background for discussions in the rest of the book.[2]

Biotechnology Revolution and the Development of Transgenic Technology

Recombinant DNA Technology

All this began in the middle of the nineteenth century when the Austrian geneticist Gregor Mendel demonstrated that many of the characteristics of a pea were passed from one generation to the next according to predictable rules, thus giving birth to a new discipline of science—genetics.[3] A century or so later, in the mid-1950s, the biologists James Watson and Francis Crick discovered the double-helix structure of deoxyribonucleic acid (DNA), the carrier of genetic information within every cell that instructs organisms to function, grow, and reproduce. As a major milestone in the history of science, the finding not only won Watson and Crick, along with Maurice Wilkins, a Nobel Prize in Physiology or Medicine in 1962 "for their discoveries concerning the molecular structure of nucleic acids and its significance for information transfer in living material"[4] but also heralded a new era of biotechnology revolution.[5] In the next decade, scientists groped their way toward putting the new knowledge into application.

In the early 1970s, Janet E. Mertz, a Ph.D. student under the supervision of Paul Berg (a professor and chair of the Biochemistry Department at Stanford University Medical Center), and Peter Lobban, another Ph.D. student in the same department, independently conceived the ideas for generating rDNA *in vitro* and using it for cloning, propagating, and expressing genes across organisms.[6] Meanwhile, Stanley N. Cohen, an assistant professor in Stanford's Department of Medicine, and Herbert W. Boyer, an associate professor of microbiology at the nearby University of California, San Francisco, first published papers describing the successful production and intracellular replication of rDNA. Simply put, rDNA technology uses restriction enzymes or chemical and physical methods as "scissors" to cut DNA fragments of interest from one organism, then uses ligase enzymes as "glues" to connect the fragments containing the desired rDNA to a target organism. The technology revolutionized the way of transferring genes with desired traits at the intracellular level.

In 1974, envisaging rDNA technology's potential of "genetically programming bacteria to produce in mass quantity fragile proteins hitherto impossible to isolate, let alone manufacture," Stanford University filed a patent application on behalf of Cohen and Boyer, who were granted the patent in 1980 after a tenacious fight within the academic and legal communities.[7] In 1976, backed by the venture capitalist Robert A. Swanson, Cohen and Boyer established the first biotechnology company, Genentech, with their propriety technology. Genentech then licensed the technology to Eli Lilly for the development and marketing of human insulin, the first mass-produced drug based on the technology, which became a watershed event in the development of biotechnology as a science-based technology.[8] The licensing also earned both inventors and Stanford an enormous amount of royalties. However, in October 1980, around the time when Genentech was listed on NASDAQ, half of the Nobel Prize in Chemistry went to Berg, the other Stanford scientist, "for his fundamental studies of the biochemistry of nucleic acids, with particular regard to recombinant DNA."[9]

Early on, some of the scientists involved in rDNA experiments felt the necessity to regulate the technology out of concern that it might be misused for unintended, undesirable, or even unsafe purposes. In January 1973, sponsored by the U.S. National Institutes of Health (NIH) and National Science Foundation, scientists gathered at the Asilomar Conference Center in California to assess the risks of working with the technology. Two years later, in February 1975, at the same Asilomar Center, an international conference on rDNA reached a consensus whereby scientists agreed to a voluntary moratorium on rDNA research until NIH formulated formal guidelines, which were put in place in the summer of 1976.[10] Since then, rDNA technology has not only become more sophisticated but has also proliferated, with applications in medicine, the environment, energy, new materials, and agriculture.

Having benefited significantly from rDNA technology, "new" genetic engineering differs from traditional genetic engineering in "the speed of the development," in the words of the physicist Freeman Dyson.[11] Applying it to crops, agricultural biotechnologists no longer need to wait for nature to come up with a desired trait. Instead, they can speed up the selective breeding process by first producing a DNA with a gene of a novel trait that could not be obtained through conventional breeding. The process requires the help of four other genes—a promoter; a terminator; a regulatory sequence of DNA that determines the location, timing, and quantity of gene expression so as to reliably produce the desired trait; and a marker whose expression signifies the successful transfer of the gene of interest to the plant's genome. The transgenic DNA can be inserted into the target organism in one of the three ways: *Agrobacterium*-mediated transformation, gene gun-mediated or particle-bombardment transformation, or pollen-tube pathway transformation. Respectively, these transformations absorb *Agrobacterium tumefaciens* with a plasmid containing the DNA to penetrate plant tissue or cell cultures; fire a "gene gun" to shoot microscopic gold particles, coated with the DNA, into a cell; or let the DNA solution be absorbed into the genome of the developing plant embryo through the pollen tube formed when the plant is in bloom or through fertilization. The gene gun-mediated transformation is used for crops such as rice and corn that are not susceptible to infection by the *Agrobacterium*-mediated transformation. In the 1980s, the Chinese biochemist Zhou Guangyu—an elder sister of Zhou Guang-zhao, former president of the Chinese Academy of Sciences—pioneered the pollen-tube pathway transformation, which has since been widely used in the transformation of cotton cultivars. Through one of the transformations, novel genetic combinations give the recipient crop traits associated with the newly inserted gene to produce a transgenic, or GM, crop.[12]

GM crops promise to deliver substantial and sustainable agronomic, socio-economic, health, and environmental benefits to farmers as well as developers, and eventually to consumers. However, first-generation GM crops were developed mainly to enhance the economics of agricultural production by offering farmers convenient field management through herbicide tolerance and insect resistance, with higher yields being a by-product.[13] An herbicide-tolerant GM crop tolerates spraying of a broad range of herbicides, mostly glyphosate, known by its brand name Roundup, developed by Monsanto, a leading agricultural biotechnology company. Roundup kills only weeds but not its Roundup Ready, or Roundup herbicide-tolerant, GM crops such as corn, soybeans, canola (rapeseed), and sugar beet, which are usually marketed together with Roundup. Similarly, an insect-resistant GM crop exudes a toxin in its pollen that kills insects that would otherwise eat the crop. Insect-resistant crops are often referred to by the name of the bacteria used, of which *Bacillus thuringiensis* (Bt), a common bacterium that exists naturally in soil, is most common. Bt gene produces a toxin that functions as a pesticide to kill Lepidopteran insects such as cotton bollworm, corn borer, and rice stem borer, or reduces the usage of pesticides, thus also having benign environmental consequences. Bt toxin is expressed only in an alkaline environment such as the digestive tract of insects but not in the acidic digestive tract of mammals. As such, it is harmless to humans and the vast majority of animals.[14] Moreover, Bt toxin functions because of the presence of a receptor in the intestinal cells of Lepidopteran insects that allows the toxin's attachment. This receptor does not exist in vertebrates, further implying that the health risk of Bt-based GM foods is not higher than that of conventional foods.[15]

Efforts have also been made to bring traits such as higher nutritional content, improved food quality including taste, and added medicinal properties to second-generation GM crops so as to benefit consumers.[16] Syngenta AG, a Swiss company (now a part of ChemChina) producing agricultural chemicals and seeds, for example, has developed Golden Rice, fortified with β-carotene, a precursor of vitamin A, to solve vitamin A deficiency that is estimated to kill 670,000 children age five and under each year.[17]

Development of GM Crops

The first GM plant is attributed to Marc Van Montagu and his collaborators at the University of Ghent in Belgium. In 1983, they implemented the first vegetal transgenesis and subsequently obtained transgenic tobacco resistant to the antibiotic kanamycin.[18] (Virus-resistant tobacco was first approved for planting in China in 1992, see chapter 3.[19]) However, it was on May 18, 1994, in the United States, that the Flavr Savr tomato made history when it became the world's first GM crop approved for human consumption. The tomato stayed ripe longer than the non-GM variety and was tastier.

Calgene, its producer, was an agricultural biotechnology start-up cofounded by Ray Valentine, a young professor in the Department of Agronomy and Range Science at the University of California, Davis.[20] The company not only successfully inhibited a gene that produces a protein making tomatoes squishy; in order to build public confidence, it also voluntarily sought government approval, labeled GM tomatoes clearly, and provided a toll-free number to answer questions from consumers. But the product was short-lived; production ceased in 1997 as mounting costs prevented it from becoming financially viable. (Calgene was later acquired by Monsanto.)

The Flavr Savr tomato's early market failure did not set back the development of GM crops. Two decades later, in the United States, 93 percent of soybeans and 88 percent of corn are genetically engineered, and GMOs are unlabeled and used in processed foods and animal feeds.[21] Globally, in 2016, 185.1 million hectares of farmland in twenty-six countries were devoted to growing GM crops, up from 1.7 million hectares in a handful of countries in 1996 (table 1.1). In twenty years, farmers in developed and increasingly in developing countries have adopted GM crops at unprecedented rates; the total number of GM-crop farmers reached 18 million in 2016. With developing countries accounting for nine of the top eleven countries growing more than one million hectares of GM crops, GM-crop growing areas reached 60 percent of the world's population.[22]

In particular, herbicide-tolerant soybeans now are being grown in the United States, Brazil, and Argentina, the three leading GM nations, as well as in Canada. Insect-resistant and herbicide-tolerant varieties of corn are widely planted in almost all leading GM nations and mainly used for feeding animals and fueling cars. As one of the earliest developers and adopters of GM crops, China, along with the United States, Brazil, Argentina, and India, has witnessed a significant percentage of its cotton variety genetically altered to contain Bt genes that resist bollworm attack. Bt cotton has been grown worldwide rapidly and widely because its commercialization is much less controversial than that of GM food crops. Meanwhile, important soybean-producing countries such as China and India have not allowed GM soybeans to be grown but instead have imported them in substantial quantities. Finally, GM rice, potatoes, and wheat, three staple crops, have been developed in the United States, but farmers have been reluctant to adopt them for fear of being blocked from the European Union and other food markets that restrict GM crops.[23] In 2015, the Innate™ Generation 1 potato with lower levels of acrylamide (a potential carcinogen) and resistance to

TABLE 1.1

Worldwide Areas Planted with GM crops (million hectares)

Year	United States	Brazil	Argentina	Canada	India	China	World total
1996	1.5	n.a.	0.1	0.1	n.a.	0.1	1.7
1997	8.1	n.a.	1.4	1.3	n.a.	0.1	10.1
1998	20.8	n.a.	4.3	2.8	n.a.	0.2	27.2
1999	28.7	n.a.	6.7	4.0	n.a.	0.3	38.7
2000	30.3	n.a.	10.0	3.0	n.a.	0.5	44.2
2001	35.7	n.a.	11.8	3.2	n.a.	1.5	52.3
2002	39.0	n.a.	13.5	3.5	<0.1	2.1	59.2
2003	42.8	3.0	13.9	4.4	0.1	2.8	67.4
2004	47.6	5.0	16.2	5.4	0.5	3.7	77.4
2005	49.8	9.4	17.1	5.8	1.3	3.3	87.2
2006	54.6	11.5	18.0	6.1	3.8	3.5	102.0
2007	57.7	15.0	19.1	7.0	6.2	3.8	114.0
2008	62.5	15.8	21.0	7.6	7.6	3.8	125.0
2009	64.0	21.4	21.3	8.2	8.4	3.7	134.0
2010	66.8	25.4	22.9	8.9	9.4	3.5	148.0
2011	69.0	39.3	23.7	10.4	20.6	3.9	160.0
2012	69.5	36.6	23.9	11.6	10.8	4.0	170.0
2013	70.1	40.3	24.4	10.8	11.0	4.2	175.0
2014	73.1	42.2	24.3	11.6	11.6	3.9	181.5
2015	70.9	44.2	24.5	11.0	11.6	3.7	179.7
2016	72.9	49.1	23.8	11.6	10.8	2.8	185.1
Major crops in 2016	Maize, soybean, cotton, canola, sugar beet, alfalfa, papaya, squash, potato	Soybean, maize, cotton	Soybean, maize, cotton	Canola, maize, soybean, sugar beet, alfalfa	Cotton	Cotton, papaya, poplar	

Note: n.a. = not available.

Source: Clive James, *Global Status of Commercialized Biotech/GM Crops* (ISAAA Annual Briefs, 1996–2015), accessed June 1, 2017, http://www.isaaa.org/resources/publications/briefs/default.asp.

bruising, developed by the American company Simplot, was first commercialized on 160 hectares; also approved was an improved version, the Innate™ Generation 2 potato, which has additional traits of resistance to fungal disease and potato late blight, the cause of the Irish famine in 1845 that led to one million deaths.[24] In Iran, commercial use of GM rice was approved in 2005, but planting was only on a very small scale.[25] In China, biosafety certificates, a prelude to commercialization, were issued for two strains of Bt rice in August 2009; they expired five years later because of hesitancy on the part of the government amid strong public opposition before being renewed in January 2015. The expected adoption of GM food crops has not yet materialized.

Controversies Around GMOs

Nowadays, biotechnology is no longer considered dangerous and is less controversial when used for biomedical purposes (red biotechnology). The "substantial equivalence" principle advocated by the U.S. Food and Drug Administration to govern risk assessments of agricultural biotechnology (green biotechnology) presents GM crops as not differing from conventional ones and therefore requiring no additional risk assessment (see chapter 2 for further discussion). Indeed, the scientific community unequivocally agrees that consuming approved GM foods is "no riskier than consuming the same foods containing ingredients from crop plants modified by conventional plant improvement techniques."[26] But concerns remain over the safety of introducing rDNA molecules and proteins into the environment or food chain. There have been widespread sensationalism and vilification that are not based upon reasoning, facts, and evidence.[27] GM crops have been portrayed as "evils" and GM foods as "Frankenfoods," a reference to the mad scientist who created a monster in Mary Shelley's famous novel *Frankenstein*.

Controversies over the effects of GM crops and foods on humans and the environment started to appear at the birth of these products, in the discourse of risk and uncertainty in which risk is seen as "calculable" and "controllable" and therefore less threatening than uncertainty, which denotes "incalculability and hence uncontrollability" and comes to be equated with "danger."[28] Seven cases seem to stand out, representing scenarios of uncertainties where GM crops and foods could possibly harm the environment or human health.

On June 22, 1998, Árpád Pusztai of the Rowett Institute in the United Kingdom appeared on television, announcing that rats fed on raw and cooked GM potatoes modified with the *Galanthus nivalis agglutinin* (*GNA*) gene from the *Galanthus* (snowdrop) plant caused thickening of the stomach mucosa, stunted growth, and suppressed immune system. His announcement resulted in frenzy. The international NGO of environmental protection, Greenpeace, and other NGOs organized rallies, calling GM potatoes "killer potatoes," destroying experimental sites and burning GM potatoes. Upon reviewing Pusztai's work, the Rowett Institute and the Royal Society of London concluded that the data did not support Pusztai's findings. But the medical journal *Lancet* nevertheless published the study in October 1999.[29]

In May 1999, a paper in *Nature* made headlines. Authored by John E. Losey, Linda S. Rayor, and Maureen E. Carter from Cornell University, the paper claimed that monarch caterpillars under laboratory settings might be harmed by pollen from Bt corn, which is supposed to produce Bt toxin to protect the corn from depredations of the European corn borer. The Cornell scientists found a higher death rate among monarch caterpillars raised on corn leaves when they were dusted with Bt-containing pollen of the corn. Because milkweed plants, the only food source for monarch caterpillars, often grow along the edges of cornfields, the scientists suspected that Bt corn might pose a danger to the butterflies. Follow-up studies organized by the U.S. Department of Agriculture (USDA) and published in the *Proceedings of the National Academy of Sciences*, however, identified a flaw in the study in that it was conducted in a confined laboratory, not in an open field. There still is no conclusive proof whether the caterpillars are at risk.[30]

In the fall of 2000, StarLink, a strain of Bt corn developed by Aventis CropScience, a subsidiary of French-German life-science giant Aventis and approved only for animal feed in the United States, was found in Taco Bell–branded taco shells sold in supermarkets. The U.S. Environmental Protection Agency (EPA) raised concerns about possible allergenicity associated with Cry9c, a Bt-gene variety. Under pressure from anti-GMO groups, Kraft Foods, producer of the taco shells, initiated a series of recalls and eventually withdrew the product from the market. The company later settled the case for $60 million with Taco Bell franchisees for lost food sales caused by damage to the Taco Bell brand.[31]

In 2001, David Quist and Ignacio H. Chapela at the University of California, Berkeley, published a paper in *Nature* about the flow of transgenes into wild Mexican maize. Controversy over the accuracy of the claims and concerns about research methodology led to an editor's note in a later issue indicating that evidence was insufficient to justify the original publication. Two independent follow-up studies in 2003 and 2004 found no evidence of transgenic DNA in Mexican maize, and GM supporters widely called for a retraction of the original paper. A study published in the February 2009 issue of *Molecular Ecology* continued to allege otherwise, although it also did not confirm that the transgene-contaminated corn had replicated.[32]

In August 2006, the USDA found that Bayer CropScience's experimental LibertyLink rice—a GM rice variety tolerant to glufosinate, the active ingredient in Liberty herbicide—in the U.S. supply of long-grain rice had contaminated two of the most popular varieties, Cheniere and Clearfield 131. The LibertyLink rice had originally been developed by AgrEvo, which was bought by Aventis CropScience, the same company caught in the Star-Link contamination mentioned above, which was subsequently acquired by Bayer AG in 2001. Shortly after USDA's announcement, Japan and Russia banned imports of long-grain rice from the United States while Mexico and the European Union required that U.S.-grown rice be tested and proven free of GM traits. According to the investigation by Bayer and the USDA, pollen drift caused contamination, which most likely occurred when field trials were carried out in 2001. In July 2011, Bayer CropScience agreed to pay up to $750 million to settle the lawsuits brought by more than 11,000 long-grain-rice farmers in Texas, Louisiana, Missouri, Arkansas, and Mississippi for their losses in rice exports.[33]

Since 2007, Gilles-Eric Séralini, a French researcher at the University of Caen, has published a number of papers claiming that GM foods present human health risks; but several of his papers have been retracted. A 2007 paper in the *Archives of Environmental Contamination and Technology* concluded that rats fed with Monsanto's MON863 Bt-corn for more than ninety days began to show "signs of toxicity" in livers and kidneys; the paper was later retracted on the grounds that Séralini did not have sufficient data to support his conclusion. In 2012, Séralini published a paper in *Food and Chemical Toxicology*, revealing further results of experiments on feeding rats over their two-year life span with Monsanto's Roundup-tolerant NK603 corn and Roundup herbicide—glyphosate—and concluding that both glyphosate and NK603 were toxic. Again, no definitive conclusions could be reached regarding the role of either NK603 or glyphosate in overall mortality or tumor rates, given the small sample size and that Sprague-Dawley rats, the type used in the study, are known for their high incidence of tumors. In addition to rampant methodological problems, this study was so ideologically biased that it was thoroughly discredited: mainstream scientists savaged it and every major independent research organization in the world reviewing it rejected it. In November 2013, *Food and Chemical Toxicology* retracted the paper after Séralini and his colleagues refused to do so. However, in June 2014, the same paper reappeared in *Environmental Sciences Europe* without a peer review.[34] Today, Séralini is a global celebrity among the anti-GMO camp and a poster child for "GMOs will kill you" and conspiracy theories about Monsanto.

Most recently, in March 2015, the International Agency for Research on Cancer (IARC), a specialized cancer agency of the World Health Organization (WHO), published an assessment of the carcinogenicity of glyphosate and four other organophosphate pesticides in *The Lancet Oncology*. The significance of the assessment is that IARC classifies glyphosate as a probable carcinogen to humans, based on sufficient evidence in experimental animals but limited evidence in humans.[35] Although Monsanto challenged IARC's conclusion, saying that glyphosate poses no human risk, the U.S. EPA is conducting a formal review of its safety.[36] As glyphosate has been widely used in the production of GM crops, GMO opponents allege that foods derived from GM crops cause cancers. But in May 2016, a joint meeting of the Food and Agriculture Organization of the United Nations and WHO on herbicide residues concluded that glyphosate poses no cancer risk.[37] And in November 2017, the European Union finally gave the green light to glyphosate for another five years despite some of its member states' campaigning to ban its usage in Europe.[38]

These are some of the most infamous cases arguably sounding real or false alarms about the possible effects of GM crops and foods on human health and the environment. They are not just about technical issues such as pollen drift or gene flow, possible effects on nontarget organisms and humans, and allergenicity reaction but have implications for risk assessment, management, and regulation pertaining to GMOs. As the introduction of GM crops or foods may have unforeseen consequences for human health or the surrounding ecosystem, and the public is particularly sensitive to possible damages caused by GMOs as opposed to those caused by conventional crops and foods, stringent biosafety and food-safety regulations, and especially their enforcement, can help promote the healthy development of their research and commercialization.

Summary and Discussion

Since the middle of the twentieth century, the life sciences have witnessed one after another breakthrough, some of which have been translated into biotechnological innovation to help transform human life in a very profound way. Accompanying the development of biotechnology, and especially the technology's application to agriculture and foods have been debates not only on the merits but also on the human health and environmental, as well as ethical, legal, and social implications of such technology. As such debates go beyond pure technology, related policymaking seems to be more complicated.

The scientific community promises that GM crops hold the key to the global food-security challenge and insists that those approved by regulators in respective countries are safe for humans and friendly to the environment. Agricultural biotechnologists also argue that the anti-GMO camp has neither the scientific knowledge and reasoning nor sufficient evidence to substantiate its positions. Having enjoyed substantial and sustainable socioeconomic and environment benefits, farmers, in increasing numbers, largely trust and are willing to grow GM crops. In May 2016, the U.S. National Academy of Sciences, Engineering, and Medicine released a new report, *Genetically Engineered Crops: Past Experience and Future Prospects.* Based on a review of 900 studies, it concludes again that there is "reasonable evidence that animals were not harmed by eating food derived from GE crops" and that epidemiological data show no increase in cancer or any other health problems as a result of these crops' entering the food supply. In particular, the report indicates that pest-resistant crops poison insects while the Bt genes generally allow farmers to use less pesticide, and farmers can manage the risk of pest resistance by using crops with high enough levels of the toxin and planting non-Bt "refuges" nearby. But the report also urges more research on strategies to delay weed resistance to glyphosate, as such resistance can "present a major agronomic problem."[39] Meanwhile, more than a hundred Nobel laureates released a letter in support of agricultural biotechnology. The letter particularly urged Greenpeace and its supporters to "re-examine the experience of farmers and consumers worldwide with GM crops and foods, recognize the findings of authoritative scientific bodies and regulatory agencies, and abandon their campaign against GMOs in general and Golden Rice in particular."[40]

But GMO skeptics contend that any form of genetic modification brings with it the possibility of unintended mutations or alternations that could pose risks to human health or the environment, thus fueling public anxiety. They maintain that clever conventional breeding, assisted by the growing knowledge of genetic markers, can achieve similar results. They also cast doubts on the motivations of agricultural biotechnology MNCs and the professionalism and integrity of members of the agricultural biotechnology community who are alleged to have their own personal interests in developing and commercializing GM crops. Moreover, given the societal implications of the technology, according to the skeptics, the debates on GMOs should not be confined to biotechnologists; policymaking related to the technology should involve other stakeholders. As will be discussed in the following pages, such sharply differing views, which seem unlikely to reach a consensus or even a compromise at least for the time being, have also dominated the discourse on transgenic technology, GMOs, and related policymaking in China.

Meanwhile, science never ceases to move forward. Recent progress in gene-editing technology, represented by CRISPR (clustered regularly interspaced short palindromic repeats)/Cas9, allows agricultural biotechnologists to use Cas9, an enzyme, guided by two strands of ribonucleic acid (RNA), to insert, delete, or replace DNA in a crop's genome in a quicker, cheaper, easier, and more efficient and precise way.[41] Genetic modification through CRISPR/Cas9, "the least biologically disruptive form of plant breeding," does not introduce any foreign gene, thereby avoiding unintended consequences and enabling scientists to sidestep, in many cases, the controversies around transgenic technology and GMOs. Perceived as a game-changing plant-editing tool of choice, CRISPR/Cas9 is expected to change the tenor of the debates over GM crops and foods.[42] It has been tested in crops, including wheat, rice, soybeans, potatoes, sorghum, oranges, and tomatoes for a spectrum of agricultural applications, from boosting crop resistance to pests to reducing the toll of livestock disease, which have the potential to produce crops based on gene-edited organisms (GEOs).[43] What remains to be seen, however, is whether the anti-GMO sentiment will turn into a new anti-GEO campaign because the new technology still involves tweaking DNAs as transgenic technology does, albeit in a fundamentally different way.[44]

CHAPTER II

Global GMO Policy

Following the initial development of recombinant DNA technology and especially foreseeing the prospect for its applications, the Organization for Economic Co-operation and Development (OECD) took three years to work out *Recombinant DNA Safety Considerations*, also known as the Blue Book, published in 1986. This policy document recommended using existing data on environmental and human health effects of living organisms to guide risk assessments; ensuring that rDNA organisms are evaluated for potential risks by an independent review on a case-by-case basis prior to their introduction into agriculture and the environment; developing rDNA organisms for agricultural or environmental applications on a stepwise basis, moving, where appropriate, from laboratory to greenhouse to limited field testing and, finally, to large-scale field testing; and encouraging further research to improve the prediction, evaluation, and monitoring of the outcome of applications of rDNA organisms.[1] The key message of the Blue Book was to base risk assessment and management of rDNA organisms on sound science. The case-by-case approach, which implied a need for individualized assessment of each case against relevant existing criteria, along with the step-by-step scaling up, signaled precaution. The document was applied to GMOs.

Some thirty years later, GMOs are present globally. Despite a scientific consensus that GM crops and foods currently on the market do not pose a risk either to human health or to the environment, OECD countries have diverged in policies toward them. This chapter focuses on how policies toward transgenic technology and GMOs have evolved in the United States, the European Union, and India. In the United States, where GM crops have been widely planted and GM foods extensively consumed since the early 1990s, the same laws and institutions that regulate non-GM crops and foods have been applied to their GM counterparts. But the European Union, where the majority of the OECD countries are located, has introduced new and separate laws and institutions for GM crops and foods. European policy is preventive in nature, whereby a GM product can be rejected simply because of "uncertainty" associated with it. The American and European policies also differ significantly in their approaches toward labeling GM foods, although consumer concerns and demands for labeling have been on the rise in the United States, leading to a GMO labeling law signed by President Barack Obama in 2016. Finally, as one of the most important agricultural countries in the world, India started to develop transgenic technology in the 1980s and subsequently established a precautious and preventative biosafety regulatory regime.

GM Policy in the United States

Formation of the Policy and Its Update

As the earliest country in the world to develop GMOs, the United States has also been the leader in the regulation of such products, following its early practice of imposing a moratorium on rDNA research in the 1970s. In 1986, the same year that OECD published its Blue Book and two years after soliciting public comments, the U.S. Office of Science and Technology Policy (OSTP) finalized a "Coordinated Framework for Regulation of Biotechnology."[2] Issued under the strongly deregulatory Reagan administration and in an atmosphere in which Americans were less worried about the environmental and health effects of biotechnology,[3] this policy document not only clarified several very important positions of the U.S. government on the management of transgenic technology and products but also laid a cornerstone for regulating GMOs in years to come.[4]

First, the framework regarded GM crops and foods as equivalent to their non–GM counterparts. Second, it proposed regulating GM crops and foods as "products" rather than treating them as a different "process." Third, because of the first two, the framework ruled out the necessity to enact new laws or establish new institutions to regulate GM products; instead, "a mosaic of existing federal law" would do the job.[5] A policy environment was thus established with the intentions, on one hand, of ensuring public safety and, on the other, of reducing overly stringent regulation that could hinder technological innovation.

In 1992, OSTP updated the 1986 Coordinated Framework. In addition to describing a risk-based, scientifically sound basis for the oversight of activities that introduce biotechnology products into the environment, the policy reaffirmed that government oversight "focuses on the characteristics of the biotechnology product and the environment into which it is being introduced, not the process by which the product is created."[6]

To address the problems in the implementation of the OSTP framework and its update, some of which are to be discussed, and the progress in bio-technology, in 2015, the Executive Office of the U.S. President Barack Obama directed the Department of Agriculture (USDA), the Environmental Protec-tion Administration (EPA), and the Food and Drug Administration (FDA) to further update the framework. Under the auspices of the National Science and Technology Council, FDA, along with OSTP and USDA, three hearings were held in 2015 and 2016 to solicit public opinions. The final document, "Modernizing the Regulatory System for Biotechnology Products: Final Ver-sion of the 2017 Update to the Coordinated Framework for the Regulation of Biotechnology," was released on January 4, 2017, just days before President Obama left the White House. But it does not represent significant changes to the previous policy and regulatory framework. In particular, the efforts are twofold. First, the update aims to "clarify the current roles and responsibilities of the agencies that regulate the products of biotechnology, develop a long-term strategy to ensure that the Federal biotechnology regulatory system is prepared for the future products of biotechnology, and commission an inde-pendent, expert analysis of the future landscape of biotechnology products." Second, the update is intended to "increase public confidence in the regula-tory system and to prevent unnecessary barriers to future innovation and competitiveness" by improving the transparency, coordination, predictability, and efficiency of the regulation of biotechnology products while continuing to protect health and the environment.[7]

Division of Responsibilities

The OSTP framework (and its updates) delegates and divides responsibility governing GM crops and foods among three federal agencies—USDA, EPA, and FDA under the Department of Health and Human Services.[8] In particular, USDA governs whether the introduction of GM crops may pose a risk to plant health and the environment as conventionally bred ones do. It derives its regulatory authority from provisions of the Plant Protection Act (PPA), which later became part of the larger Agriculture Risk Protection Act of 2000 (known as ARPA 2000) under the Clinton administration. Using a permission and notification system, the complex deregulation process begins when an organization wishes to import, move interstate, or field-test a GM crop. After several years of field testing and data collection, the organization that wants to commercialize the GM crop needs to petition USDA's Animal Plant Health Inspection Service (APHIS) for a determination of nonregulated status. APHIS then assesses the plant in accordance with both PPA and the National Environmental Policy Act (NEPA) to ascertain that the new crop variety is not a pest, poses no threat to agriculture or the environment, and therefore is no longer regulated by USDA.[9]

The Federal Insecticide, Fungicide, and Rodenticide Act (FIFRA) and the Federal Food, Drug, and Cosmetics Act (FFDCA) guide EPA's regulation of two types of pesticides associated with GMOs: plant-incorporated protectants (PIPs) and genetically modified microbial pesticides (GMMPs). PIPs are created by transferring specific genetic materials from a bacterium to a plant, while GMMPs are either bacteria, fungi, viruses, protozoa, or algae whose DNA has been modified to express pesticidal properties in a living plant. EPA also pays close attention to the potential pest resistance to Bt-based PIPs.

Finally, FDA regulates premarket approval of all food products by holding their producers accountable for the products' safety and proper labeling. In 1992, FDA issued a "Statement of Policy: Food Derived from New Plant Varieties" to clarify its interpretation of applying FFDCA to human foods and animal feeds derived from new plant varieties and guiding the industry on related scientific and regulatory issues.[10] In particular, the policy proposed treating foods developed using agricultural biotechnology as "substantially equivalent" to conventional foods and, as such, "generally recognized as safe"—a position that was also endorsed by the U.S. National Academy of Sciences (NAS) and recently the U.S. National Academies of Sciences, Engineering, and Medicine.[11]

According to FDA policy and additional guidance issued in 1996, a developer intending to commercialize a GM food should meet with FDA to identify and discuss the food's relevant safety, nutritional, or regulatory issues before submitting a summary of the food's scientific and regulatory assessment to the agency. FDA then evaluates the submission and typically responds by letter, again reminding the developer of its responsibility to ensure the food's safety. In 2001, FDA issued a "Premarket Notice Concerning Bioengineered Food." While reaffirming the "substantial equivalence" position, the agency further required that a developer submit a scientific and regulatory assessment 120 days before marketing GM foods and recommended that the developer continue consultation with FDA before submitting the premarket notice and seek approval in case "the genetic manipulation of the seeds used to produce them had altered their substance or safety, in which case they would be regulated as food additives."[12]

Of the three federal agencies, USDA, because of its long history and positive reputation, emerged as America's primary agency regulating the introduction of GM plant varieties into the environment for either field tests or commercial production, with the recently created EPA playing a secondary role. For example, regarding herbicide-tolerant, most commonly glyphosate-tolerant, crops, USDA regulates the crops but EPA regulates the herbicides.[13] The "substantial equivalence" position has also facilitated an autonomous and centralized environment in which regulators take science- and evidence-based approaches to simultaneously make and implement policy and rapidly deregulate GMOs. Afterwards, GM crops may be treated in the same manner as their parent plants, grown and marketed following normal food-safety regulations.

The three agencies have followed the regulatory framework to approve GM crops for agricultural production and the foods and feeds from GM crops for human and animal consumption. Since 1992, when Flavr Savr tomato was approved, USDA has approved a total of 125 petitions for commercial release (deregulation) of transgenic traits in nine different major crops—corn (maize), soybean, cotton, canola, sugar beet, alfalfa, papaya, squash, and potato—the latest one being canola with male sterility and glufosinate-ammonium tolerance, developed by Bayer, on July 26, 2017.[14] FDA has completed 155 consultations, mostly on corn; the latest consultation was on October 20, 2017, for the same Bayer canola recently deregulated at USDA.[15] And since 1995, EPA had registered twelve PIPs and eight GMMPs.[16] Having always been the world's largest GM nation, the United States planted 72.9 million hectares of GM crops in 2016 (see table 1.1).

Criticism of the Policy and New Development

Both the pro- and anti-GMO camps in the United States have criticized the nation's GMO policy. First, a simplified regulatory regime does not mean an easy process for developers, as USDA, EPA, and FDA have all invoked triggers for mandatory or voluntary regulatory review on GM crops. For example, USDA has created a new category of "regulated article" for field testing and commercial release of transgenic crops; EPA has established new categories of PIPs and GMMPs; and FDA has adopted a policy of voluntary consultation for foods containing GM ingredients. Each regulator thus subjects GM crops or foods to additional and sometimes demanding regulatory oversight. In fact, compared with conventional crops, the deregulation process for a GM crop has become more "time-consuming, costly and effort-intensive," requiring five to seven years, many millions of dollars, much greater documentation, and large amounts of data. More discouraging is the fact that some GM products have been "in regulatory limbo without any decision other than additional requests for information and time to study."[17]

Second, U.S. GMO regulation creates litigation risks. For example, PPA authorizes USDA to restrict the introduction of organisms that might harm plants. Technically, GM crops fall into the category of "plant pest" organisms because they use DNA promoters derived from natural plant pathogens, most notably cauliflower mosaic virus, to amplify the introduced traits; NEPA stipulates that USDA assess the environmental impact of novel crops under regulation; and the Endangered Species Act (ESA) also requires that USDA gauge potential impacts of GM crops on endangered species. These laws and statutes have given GMO activists reasons to challenge decisions by the relevant agencies, sue them, and force rigorous supervision. In recent years, the Center for Food Safety, a pro-GMO-labeling public-interest and environmental advocacy, sued USDA for failure to conduct proper environmental-impact assessments and endangered-species analyses for crops in deregulation.[18] Regardless of the outcome, these lawsuits have portended delays, incurred litigation expenses, and generated great uncertainties upon GM products. In addition, because of the strict regulatory oversight, any GM trait that is found in unapproved crops or uses could immediately evoke class-action lawsuits for economic damages.[19] The lawsuits against StarLink and LibertyLink contaminations, mentioned in the last chapter, are such examples. Therefore, in order to market its product, a GM developer needs to successfully and simultaneously survive both the regulatory process for the product and the judicial system.[20]

Third, assessing GM crops that produce their own pesticides leaves EPA with "a limited role in overseeing the environmental impacts of GMOs." Since 1986, the agency has complained regularly about this with modest success.[21]

Debates on Labeling

Most noticeable of the American GMO activism has been its increasingly harsh stance on the issue of labeling.[22] Until July 2016, there existed neither any federal labeling requirement on GMO thresholds nor mandatory GMO labeling as FDA's "substantial equivalence" position categorically denied any possibility of labeling GMOs.[23] The food industry also resisted labeling GM foods, claiming that singling out GM ingredients might stir unnecessary and unwarranted fear over their safety.

But there have been rising calls for labeling GM foods. In 1999, two bills on labeling of GM foods were introduced, unsuccessfully, in the Senate and House. Also in 1999, when it held public hearings on GM foods, FDA received more than 50,000 written comments with the majority supporting the labeling of such foods. However, citing FFDCA, FDA firmly rejected mandatory labeling. It also was specifically against using the misleading term "GMO-free" as all food is likely to contain some GM ingredients.[24] NGOs have filed lawsuits, unsuccessfully, against FDA and food companies on labeling. Recently, public opinion seems to have become overwhelmingly in favor of disclosure. A *New York Times* poll in July 2013, for example, found that 93 percent of Americans wanted GM foods labeled.[25]

GMO labeling has become a battleground at the state level. Vermont, Connecticut, and Maine have passed GMO labeling bills,[26] but similar state ballot initiatives in California, Washington, Colorado, and Oregon have repeatedly lost;[27] the defeat in Oregon in 2014 was the second one, following a similar defeat in 2002.[28] According to the Center for Food Safety, more than seventy bills and ballot initiatives have been introduced in more than thirty states, most noticeably in Massachusetts, New York, and New Hampshire, requiring labeling or banning GMOs completely.[29] It cost the food and agricultural industry more than $70 million to defeat the mandatory GMO labeling initiatives in California and Washington in 2012 and 2013; Monsanto alone spent more than $8 million in California and more than $5 million in Washington.[30]

But opponents to mandatory GMO labeling did not want to keep waging multi-million-dollar, state-by-state legal fights. While supporting the consumer's right to know, Monsanto blamed labeling supporters for trying to create false impressions and mislead consumers that GM foods are harmful and unsafe. Perceiving that state-by-state labeling fights could be costly for food manufacturers and distributors, the U.S. Grocery Manufacturers Association (GMA), an organization of more than 300 food companies, funded efforts in twenty-five states to counter labeling measures and pushed for changes at the federal level in the oversight and labeling of GM foods. On one hand, GMA formed a broad coalition with GM developers and other stakeholders to propose legislation that would make all GMO labeling voluntary. Such legislation aimed at specifically preempting any state labeling regulation different from the federal one and hedging against a vocal consumer movement that had been making inroads at the state level. On the other hand, GMA proposed making it mandatory that GM developers notify FDA before marketing new products so as to give consumers greater confidence and also requiring that FDA respond to mandatory notifications within specific time frames. Such legislation would not change FDA's oversight but would allow it to ascertain that "any special labeling" aims to protect health and safety and to "prevent the label of bioengineered food from being false or misleading."[31]

GMA's efforts finally led to passage of H.R. 1599, the Safe and Accurate Food Labeling Act, by the House of Representatives on July 23, 2015.[32] Almost a year later, on July 7, 2016, the Senate passed its version of the GMO-labeling bill, which, with industrial backing, would require that food and beverage companies disclose GMO ingredients in their products. On July 14, a similar bill was passed in the House. Two weeks later, on July 28, President Barack Obama signed the bill into a law, making it override stronger state legislations such as Vermont's, which had gone into effect on July 1.

It is worth noting that the ingredient information is not to be put directly on the package but provided indirectly through a QR code, a type of bar-code, or obtained by calling the producer or visiting its website, thus causing criticism by the pro-GMO-labeling activists.[33]

Alongside these recent legislative developments have been moves by food retailers. In spring 2013, Whole Foods Market, an organic-food supermarket chain, announced that it would label all GM products sold in its U.S. and Canadian stores by 2018; Chipotle Mexican Grill, a Mexican-food restaurant chain, has been working to eliminate GM-derived foods from its menu;[34] and Ben and Jerry's, which campaigned in support of Washington State's labeling initiative, planned to begin producing GMO-free ice cream by 2015.[35] In early 2014, General Mills, which had been behind the efforts to defeat GM-labeling campaigns in several states, started to source non-GMO cornstarch and sugar. Symbolic and applying only to Cheerios, its original breakfast cereal, because purging GM ingredients from other types of cereals such as Honey Nut and Apple Cinnamon would be difficult, the change was hailed as "a huge victory for the non-GMO movement and consumers." According to GMO Inside and Green America, two activist groups, General Mills mainly responded to pressure to make the change, which Tom Forsythe, a General Mills spokesman, rebutted: "It's not about safety. And it was never about pressure. . . . We did it because we think consumers might embrace it." However, the change did not make Cheerios more competitive; Americans do not buy more Cheerios simply because they are GMO-free and labeled as such.[36]

GM Policy in the EU

Initiation of the Policy

The European Union (EU) is known for its strong preventive stance on GMOs, based primarily on concerns over their possible health and environmental risks. Since 1990, the European Commission (EC), EU's legislator, and some EU member states have created new laws and regulations and new regulatory pathways applicable only to GM crops and foods.

Of the various factors that galvanized the establishing and strengthening of laws and regulations governing GMOs in the EU, the first was a series of food-safety scandals. In 1986, bovine spongiform encephalopathy (BSE), also known as mad cow disease, broke out in the United Kingdom, but the meat of infected cows continued to be sold for human consumption through 1988. Moreover, the British government initially assured the public that there was no scientific evidence to link BSE in cattle and Creutzfeldt-Jacob Disease (CJD), a rare and fatal disease damaging human brain. But in March 1996, when ten British citizens were known to be infected with CJD, the government dramatically reversed its earlier position by acknowledging a possible causal link between BSE and CJD. This was "the biggest crisis the EU had ever had," said Franz Fischler, European Commissioner for Agriculture, and it "severely damaged the credibility of the EU authorities as regulators and risk managers."[37] The scandal also caused rising distrust of science and experts.[38] The BSE/CJD crisis was followed by other food-safety scandals, such as contamination of farm animal feed with dioxin, a carcinogen, and contamination of carbon dioxide with traces of chemicals that were injected into Coca-Cola at a bottling facility. The failure of European regulators to react to and manage these crises not only shocked the public but also highly politicized the food-safety concern.

Coincidently, in the fall of 1996, shipments of GM soybeans and corn arrived from the United States, spurring a widespread public outcry. Thus, the decision to reject the import of GM soybeans and corn was not scientifically driven; rather, it was due to the increasing resistance from the public and consumer-rights and environmentalist groups that linked both food-safety scandals and GM crops with "consequences of industrialized agriculture."[39] Triggered by these developments, the EU began to see the emergence of a kind of "precautionary risk culture" and to implement "progressively more stringent regulatory requirements" on GMOs.[40]

Core of the Policy

The EU's policy on GMOs was introduced as a series of directives and regulations, covering contained use, deliberate release to an outside and unconfined environment, and traceability and labeling.[41] The policy started with defining GM foods as "novel," or "unnatural," ones that are not yet used for human consumption so that regulators lack scientific evidence and definitive proof of their safety. All "novel" foods—GM or not—are regulated identically and uniformly, subject to new scientific research to assess their risks in the member state where the product would first be introduced. This "novel food" definition marked a sea change in the EU's policy toward GM foods as if they were created by a completely new and different "process," thus requiring additional precautionary measures to examine both actual risks and speculative hazards to ensure "no risk" before granting approval and to monitor them after approval for release into the environment or the market. A "novel" GM food would only be allowed to enter the EU market upon individual and explicit risk assessments and regulatory approval to prove its safety. Moreover, once "novel," forever "novel"; there is no pathway for a "novel" food to become "normal."

The policy also imposed additional labeling and traceability requirements on GM foods. First, it set a 0.9 percent threshold of GM content for labeling, which would have the de facto effect of warning customers. Second, it made the traceability of GM foods an important measure in the risk management of the foods. If there were an issue or question regarding them, the labeling and traceability requirements would make them more easily identifiable and traceable so that the products could be immediately pulled out of the market.

In January 2000, the EC released a white paper outlining its position on the future of European food-safety regulation. The document applied the precautionary principle on GMOs, with the purpose of quantifying risk and minimizing hazards before full burden of proof is established. It discussed the need to separate risk assessment and management, which would allow a GM food approved by one agency to be regulated later by another. The white paper created more regulatory accountability and introduced measures to prevent "novel" foods, including GM foods, from being approved. It also proposed the establishment of a single independent European Food Safety Authority (EFSA) as the most appropriate response to the need to guarantee a high level of food safety and to assess environmental consequences of novel foods, which was reaffirmed in 2002.[42]

As European attitudes toward "novel" foods could be country- and culture-specific, implementation of the EU regulation at the member-state level would not necessarily be smooth. Consequently, discrepancies in risk assessment might result in some nations approving GM foods while others did not. Therefore, the white paper tried to standardize measures on GM products across EU member states.[43] The adaptation of these measures and their harmonization into existing legislative and regulatory frameworks in EU member states have effectively reduced the latitude for noncompliance.

Accompanying the strengthening of the complex and intricate GMO regulatory regime was an EU-wide de facto moratorium on commercialization of GM crops in June 1999. It is in this sense that EU governance on GMOs was viewed as a "supranational Leviathan."[44]

A Paradoxical and Dysfunctional Policy

Criticized as paradoxical, dysfunctional, and "based on political expediency and short-term economic goals rather than rational scientific evidence and long-term economic models,"[45] the EU policy on GMOs suffers from three major problems. First, on the surface, the policy seems to support the coexistence of transgenic and conventional crops; in reality, European farmers who want to grow GM crops would face harsh regulations that vary nation by nation. In Latvia, for example, a farmer must leave four kilometers of space between GM and non-GM canola-oil-producing rapeseed, or six kilometers if the non-GM rapeseed is certified as organic. This would effectively prevent farmers from planting GM crops; unless they agreed to surround GM crops with large areas of uncultivated land, they would risk litigation from surrounding farmers.

Second, while imposing an EU-wide de facto moratorium on growing GM corn and soybeans, the EU has allowed the imports of the exact same GM crops in fairly large quantities from the United States, Brazil, Argentina, and Canada for use as foods and feeds. Because such products have to be labeled, the United States and Argentina have sued EU at the World Trade Organization on the ground that GMO labeling is a trade barrier.[46]

Third, the de facto ban on GM crops not only has been undermining EU's competitiveness in agriculture but also may have enlarged its gap with the United States in agricultural biotechnology.[47] Indeed, the ban has adversely affected BASF and Syngenta, two Europe-based agricultural biotechnology giants. In early 2012, BASF literally abandoned Europe for the United States and Brazil, where the environment for approval and public acceptance of GM products is more favorable. Moreover, Europe risks lagging behind in pharmaceutical research. Edible vaccines, compounds to prevent HIV transmission, and insulin are just a few of the pharmaceutical ingredients that scientists have tried to integrate into GM crops. For example, GSK Biologicals, a global biotechnology giant headquartered near Brussels, Belgium, is working on this line of research and development.[48]

In fact, the Eurobarometer survey in 2010 showed mixed attitudes toward technology and innovation among Europeans. On one hand, the majority of those surveyed remained optimistic about biotechnology and genetic engineering and did not reject the impetus toward innovation and commercialization. On the other hand, they were in favor of regulation to balance market forces, wanting to be involved in policymaking on new technologies when social values are at stake. They also distrusted science, business, and regulators.[49] In no European country did a majority of citizens favor GM foods, ranging from 44 percent in the United Kingdom to only 10 percent in Greece and Cyprus. In general, concerns about safety remained paramount, followed by the absence of apparent benefits to consumers.

However, labeling did not necessarily turn consumers away from GM foods. According to another survey, three in four European consumers claimed to be aware that GM foods have to be labeled, but two-thirds of those could not distinguish GM from non-GM foods. This meant that either fewer than 50 percent of the consumers bothered to check labels in buying a food item, the information on the labels was misunderstood or misinterpreted, or consumers simply ignored labeling. Only 30 percent of survey respondents chose to avoid buying foods with GM ingredients. Most consumers were neither really interested in nor very attentive to the presence of GM foods; perhaps they did not know what they had bought or whether they had bought GM foods. Finally, though half of the survey respondents indicated that they had not bought GM foods, the barcode analysis of their purchases showed that half of those had in fact bought such products. In short, most consumers did not actively avoid GM foods, nor were they greatly concerned with the GMO issue.[50]

The fact that GM-labeled foods were available and actually bought also showed the existence of a market for GMOs. This market might be even larger than believed, as 20 percent of the non-GMO buyers thought they were already buying GM foods, and around 30 percent did not know whether they were doing so. Nonetheless, after labeling regulations were introduced, large food manufacturers and suppliers became reluctant to produce and carry GM foods, leading to a significant decline in the number of such foods on the European market and restricting the choice of European consumers to purchase GM foods.[51]

With strong political support from the EC, environment NGOs and member-state governments antagonistic to GMOs have dominated the policy debates and the media, as well as the minds of consumers. Activists have vandalized GM-crop-testing fields or sued GM-food producers. While their efforts have helped to heighten resistance to GM crops and foods, scientists have moved away from undertaking research in agricultural biotechnology and retailers have purposefully avoided GM products. In light of the EU's complicated regulatory process, particularly the importance given to risk assessment, approval of any GM crop has stalled with more than a dozen GM crops stuck somewhere in the regulatory pipeline, because of either an absence of support from a majority of member states or a member state's failure to submit a case to the EC for approval.

Consequently, although the EC has allowed quite a number of GM crops to be imported as foods and feeds, only a small number of GM-labeled products have been available in a few European countries. In fact, Europe is the only continent that has experienced a decline in the areas growing GM crops. In 2016, the entire European acreage of GM crops was 132,000 hectares, negligible in the 185.1 million hectares of GM crops planted globally. Bt corn, based on Monsanto's MON810, is grown mostly in Spain, the only European country in the top fifteen GM nations.[52] In 2013, the EU General Court reversed the approval of BASF's Amflora potato, a variety of high-starch potato and the only other GM crop allowed to be grown in Europe, although the company launched a legal battle to reverse the decision.[53]

The European scientific community has become more vocal, claiming that the opposition to GM crops is based more on fear than on science and evidence. Having spent €300 million on research on the biosafety of GMOs since 1982, in 2010 the EU concluded that "biotechnology, and in particular GMOs, are not per se more risky than e.g. conventional plant breeding technologies."[54] But as the British microbiologist John E. Beringer stated, "the anti-GM lobby's agenda has hardly changed at all. They profess still to be concerned that we have insufficient knowledge and that no GMO should be released until we can predict with certainty what it, and the cloned genes within it, will do in the environment."[55]

Leading European scientists have also voiced their concerns. In July 2012, EC's chief scientific adviser Anne Glover pointed out that "there is no substantiated case of any adverse impact [of GM foods] on human health, animal health or environmental health, so that's pretty robust evidence, and I would be confident in saying that there is no more risk in eating GMO food than eating conventionally farmed food."[56] But in November 2014, she was removed from the position, presumably because of her pro-GMO stance.[57] In 2013, in an editorial in *Science*, Louise O. Fresco, a university professor at the University of Amsterdam who would become president of the Wageningen University and Research in the Netherlands on July 1, 2014, and is also a former United Nations assistant director general for agriculture, pointed out:

> Europeans tend to romanticize the pre-modern past, unaware of the suffering and food scarcity associated with low crop yields. . . . Europe's lack of trust in GMOs reflects a wider distrust of science. . . . Only political courage, as shown last year by the British government's request for the EU to make it easier to grow GMOs, can break the ideological stalemate between NGOs, producers, consumers, and scientists.[58]

Brian Heap, president of the Science Advisory Council of the European Academies, also argued that Europe should rethink its position on GM crops as second-generation GM crops avoid some of the problems that previously provoked hostility.[59]

In fact, the EU has recently seen some new developments. In 2012, the British government requested that the EU loosen its policy on growing GMOs; EFSA fast-tracked approval of Monsanto's Intacta GM soybeans imports used as foods and feeds and declared that there is no requirement for research or allergy testing as they are "substantially equivalent to the natural bean." In 2014, EFSA rejected a French ban on GM corn.[60] And it has taken firm stands on science. For example, EFSA ruled that the French researcher Gilles-Eric Séralini did not provide credible data to support his conclusion on the relationship between GM corn and cancer in lab rats;[61] it discounted International Agency for Research on Cancer's view on glyphosate by suggesting that the weed-killer is unlikely to cause cancer, an opinion that led the EU to renew approval for glyphosate use in its member states.[62] The most recent policy allowing an individual member state to decide whether to block the cultivation of a particular GM crop on its own territory has been hailed as an attempt to break the deadlock.[63] Amid these were two megadeals, both announced in 2016, involving European agricultural biotechnology MNCs. The Swiss-based Syngenta was to be acquired by China National Chemical Corp (ChemChina) for $43 billion; ChemChina would like to maximize the potential of Syngenta's seed business, including GM-seed business, while helping grow China's GMO-centered agricultural biotechnology.[64] The Germany-based Bayer's $66 billion takeover of Monsanto represents the irony that may redraw the global GM landscape and fundamentally change Europeans' attitude toward GMOs.[65]

Finally, in a January 2013 speech to an Oxford farming conference, Mark Lynas, a Briton who was once a leading GMO critic, apologized for "having spent several years ripping up GM crops."[66] He recalled his initial negative reaction toward GMOs and Monsanto:

When I first heard about Monsanto's GM soya I knew exactly what I thought. Here was a big American corporation with a nasty track record, putting something new and experimental into our food without telling us. Mixing genes between species seemed to be about as unnatural as you can get—here was humankind acquiring too much technological power; something was bound to go horribly wrong. These genes would spread like some kind of living pollution. It was the stuff of nightmares.[67]

In fact, Lynas first wrote about "evil" GMOs and Monsanto in *Corporate Watch*, a magazine that he cofounded. Since the mid-1990s, he had helped to start "the most successful campaign," not only making ideological arguments and converting them into civil disobedience, destroying experimental GM crops and attacking agricultural biotechnologists, but also effectively helping achieve GM bans in Europe, Africa, and Asia. In 2008, he claimed, "the technology moves entirely in the wrong direction, intensifying human technologies and manipulation of nature, when we should be aiming at a more holistic ecological approach."[68] But gradually Lynas found it difficult to reconcile his environmentalism on GMOs with that on climate change, being anti-science on the former but pro-science on the latter. He also realized that the need to feed a growing world population and an exploding middle class demanding more and better-quality foods makes it necessary to take advantage of all available technologies, including transgenic technology.

Feeling bad about his erstwhile activities, Lynas eventually recanted his stance, relinquishing his leadership role in the anti-GMO movement.[69] His dramatic conversion aliened GMO activists, who labeled him "one of the biotech industry's most sought after 'ambassadors' (i.e. undercover spokespeople)."[70]

GM Policy in India

An Elaborate and Rigid Biosafety Regime

In the 1980s, India began to embrace biotechnology and its potential for agriculture. Now, more than twenty crops are at different stages of development, including foods and economic crops with pest resistance, herbicide tolerance, nutritional enhancement, and abiotic stress (e.g., drought, salinity, and soil nutrients) tolerance. Meanwhile, India has established an elaborate and rigid biosafety regime, designed around the Environmental Protection Act of 1986 (EPA 1986) and the Rules for the Manufacture, Use/Import/ Export and Storage of Hazardous Microorganisms/Genetically Engineered Organisms or Cell, 1989 (Biotechnology Rules 1989). Together, EPA 1986 and Biotechnology Rules 1989 govern a chain of activities from R&D, large-scale use, and imports of GMOs and their products through four federal government agencies and state governments involving representatives from government, the research community, and various interest groups.[71]

In particular, the Ministry of Environment and Forests (MOEF), through a Genetic Engineering Approval Committee (GEAC), implements Biotechnology Rules 1989. Accordingly, all GMOs or GMO-derived products must receive formal approval from GEAC prior to commercialization and imports. In 2008, GEAC also introduced "Guidelines and Standard Operating Procedures for the Conduct of Confined Field Trials" and "Guidelines for Safety Assessment of Foods Derived from Genetically Engineering Plants."

The Department of Biotechnology (DBT) under the Ministry of Science and Technology (MOST) provides guidelines and technical support to GEAC, and evaluates and approves biosafety assessment of R&D on GMOs. In 1990, DBT formulated "Recombinant DNA Guidelines," which were amended in 1994. In 1998, it issued separate guidelines for GM plant research, including regulations on the import and shipment of GM plants for use in research. In 2008, with experience in overseeing the development of biotechnology, DBT for the first time came up with the idea of enacting a National Biotechnology Regulatory Authority Act. In 2012, DBT submitted a Biotechnology Regulatory Authority of India Bill (BRAIB) to Parliament, which subsequently referred the bill to its Standing Committee on Science, Technology, Environment and Forests for review and stakeholder consultations. BRAIB specifically proposed setting up an independent, autonomous national regulatory authority for biosafety clearance of GM products and processes, thus transferring the ultimate authority for approving the introduction of GM crops from GEAC under MOEF to DBT under MOST. But the bill was not debated in 2013 and lapsed in 2014 because of India's national election; it has not been reintroduced in Parliament by the current National Democratic Alliance (NDA) government.[72]

In 2006, upon enacting the Food Safety and Standards Act, India established the Food Safety and Standard Authority (FSSA) under the Ministry of Health and Family Welfare. FSSA evaluates and approves the safety assessment of GM foods, including processed foods, for human consumption. On August 23, 2007, however, MOEF announced that processed foods derived from GMOs do not require approval from GEAC for production, marketing, import, and use in India because the end products are not live modified organisms (LMOs); processed foods were not even considered an environmental safety concern under EPA 1986 because they are not replicated in the environment.[73] Finally, the Ministry of Agriculture (MOA) is charged with the evaluation and approval of the commercial release of GM-crop varieties after field testing that assess agronomic performance.

At the state level, relevant government agencies monitor safety measures at biotechnology research facilities, assess damages, if any, caused by the release of GMOs, and approve field tests and commercial cultivation of GM crops that have been approved by GEAC. DBT, MOA, and state governments also support R&D on agricultural biotechnology at public R&D institutes and universities.[74]

Since 2002, the Indian government has approved six Bt-cotton events and some 1,400 Bt-cotton varieties and hybrids, mostly derived from two Monsanto events—MON531 and MON15985—as well as domestically developed events and an event from China for commercial cultivation in different agriculture-climate zones. In 2016, 10.8 million hectares were estimated to grow Bt cotton, compared to 7.6 million hectares in 2002. This represents more than 96 percent of India's cotton-producing area, making

India the world's largest Bt-cotton producer.[75] Bt cotton has been approved for seed, fiber, and feed production/consumption and has been exported as a fiber product (cellulose) and occasionally as seeds and meals.

Estimated at $5.2 billion, 15 percent of the total, in the fiscal year 2015/16 (April/March), Bt cotton has turned agricultural biotechnology into the third largest pillar of India's biotechnology industry, after pharmaceuticals and services.[76] With Bt cotton being the only GM product approved and the Bt-cotton acreage nearly at its maximum, growth of agricultural bio-technology had slowed from 15 percent in 2011/12 to 5 percent in 2012/13 and 4.3 percent in 2013/14, and is likely to slow further for the foreseeable future.[77] India has imported a large quantity of oil derived from glyphosate-tolerant soybeans and a small quantity of herbicide-tolerant canola oil. In 2010, soybean oil imported from the United States alone reached a record of $132.9 million; but this trade has also declined significantly and reached only $28.3 million in 2013. In 2015, GM soybeans were mostly imported from Argentina, Brazil, and Paraguay, while GM canola oil was imported from Canada.[78]

On February 9, 2010, MOEF imposed a moratorium on Bt brinjal, also known as eggplant or aubergine. The move concerned the agricultural biotechnology industry over a breakdown of India's biosafety regulatory regime. On July 6, 2011, GEAC introduced new procedures for authorizing GM-crop field tests, requiring developers to obtain a "no objection certifi-cate (NOC)" from relevant state governments. Only a few states have issued NOCs. In April 2012, GEAC stopped approving new GM event and field-testing applications, some of which had already gone through the requisite regulatory procedures. The last GEAC tenure ended on June 9, 2012, and it took MOEF more than nine months to form a new GEAC. The new committee met on March 22, 2013; GEAC posted decisions on its website almost three months later, on June 18, but withdrew them two days later because MOEF claimed that it had not approved them.[79] GEAC did not reconvene until March 2014. On July 17, 2014, two months after the new NDA government took power, GEAC restarted its business by granting field tests of several GM-crop events, but its action was opposed by several ideological organizations affiliated with the government.[80]

Finally, in May 2012, in response to a petition filed in 2005 that alleged that field tests of GM crops were being allowed without proper biosafety assess-ments, India's Supreme Court appointed a six-member Technical Expert Committee to review and recommend biosafety risk assessment studies for GM crops. On July 18, 2013, the committee submitted its final report, with one dissent, recommending a ban on all GM-crop field tests until the gaps in the existing biosafety regulatory regime had been addressed. The Indian government and the agricultural biotechnology industry strongly contested the recommendation immediately and during subsequent hearings in August 2013 and April and May 2014. The case remains pending with the Supreme Court.[81]

While recognizing that in due course "we must make use of genetic engineering technologies to increase the productivity of our agriculture" because of the technologies' enormous potential, in 2012, Prime Minister Manmohan Singh, in an interview with *Science*, was cautious toward the technology. Indeed, in 2009, he had not intervened when the environment minister rejected a scientific panel's advice and banned commercial planting of Bt brinjal pending additional safety tests.[82] By contrast, the current prime minister, Narendra Modi, and other senior government officials seem to be more supportive of the development of agricultural biotechnology.[83] In 2016, substantial progress was achieved toward approving a GM mustard variety developed by Delhi University. GEAC is currently reviewing the public comments. It has also established a technical subcommittee to review the safety of GM mustard for environmental release. According to MOEF, at least two or three other GM-crop events could be ready for approval.[84]

The Case of Bt Brinjal

In 2005, India's leading seed company—Maharashtra Hybrid Seeds Company, also known as Mahyco, in the West India state of Maharashtra—started to collaborate with the University of Agricultural Sciences and Tamil Nadu Agricultural University to develop Bt brinjal. By using *Agrobacterium*-mediated transformation, scientists inserted Cry1Ac, a Bt gene obtained from Monsanto, along with other genetic materials, into the genome of traditional cultivars of brinjal, making it effectively resistant to Lepidopteron insects in general and the brinjal fruit and shoot borer (*Leucinodes orbonalis*) in particular.

In 2006, after examining the biosafety data presented by Mahyco, an Indian expert committee determined that Bt brinjal had clearly demonstrated benefits such as increased yield, reduced use of pesticides, farm safety, labor productivity, farm income, product quality, and product pesticide-residue safety. But the committee also concluded that while the current data showed that Bt brinjal was safe and equivalent to its non-Bt counterpart, more study was still needed to confirm the findings and the benefits from growing Bt brinjal over existing methods for pest management and pesticide reduction. It also recommended large-scale tests. In 2009, having examined the data from these further tests, a second expert committee concluded that the benefits of Bt brinjal far outweighed the perceived and projected risks. Therefore, it advised GEAC to recommend approval for Bt brinjal's commercialization.

GEAC's decision, on October 14, 2009, prompted concerns among some scientists, farmers, and especially GMO activists. The activists, through the media, spread misinformation and disinformation on issues that emotionally influenced the public. For example, one sensational argument was that the introduction of Bt brinjal would jeopardize the use of brinjal in indigenous medicine.[85]

On February 2, 2010, after an apparent public outcry and rounds of debates involving various stakeholders, Jairam Ramesh, then environment minister, denied the scientific regulatory process by rejecting GEAC's recommendation. He said that the ministry still needed more time before approving commercialization of Bt brinjal and that there was no hurry to grow Bt brinjal in India, thus facilitating a de facto moratorium on its commercialization until further, unspecified, tests are conducted. Two weeks later, while reiterating that the ban was temporary, Minister Ramesh indicated that Bt brinjal would be released only after "we arrive at a political, scientific and societal consensus." He also alleged that Monsanto controlled 90 percent of the GM seeds in India, which was baseless, as the company's role was limited to providing the Cry1Ac gene used in Bt cotton and Bt brinjal.[86] Furthermore, in 2010, MOEF downplayed the role of the Genetic Engineering Approval Committee by changing its name to Genetic Engineering Appraisal Committee, although the acronym—GEAC—is the same.

In a generally GMO-unfriendly environment, India has seen widespread bio-piracy, or misappropriation, of local germplasm. The irony is that in India, any GM crop can be planted as long as its developer hides the fact that the crop is transgenic. While saving farmers enormous amount of time and money, growing GM crops this way bypasses the regulators, which could jeopardize biosafety and the environment. Unconfirmed cases have been reported of illegal cultivation of Bt brinjal as well as pest-resistant and herbicide-tolerant gene-stacked cotton and virus-resistant papaya in some parts of the country.[87] As of February 2010, at least ten Bt-brinjal varieties were available in Karnataka and Tamil Nadu without prior approval from the central and state governments, according to Environment Support Group, a charitable trust in Bengaluru.[88]

Bt Cotton and Suicide

The rapid adoption of Bt cotton in India and the possibility that India would introduce other GM crops have engendered mixed public opinions. Domestic activists have evoked ideological imperialism or prejudice against Western countries and MNCs, especially Monsanto, to fuel opposition. They have petitioned India's Supreme Court, vandalized R&D institutes and GM crops in officially approved test fields, and mounted campaigns to undermine transgenic technology and to pressure the Indian government to ban GM crops completely. One of the most infamous activists is Vandana Shiva.

Since the 1990s, Shiva, a physical scientist turned activist, has caught the world's attention by alleging that Maharashtra is "India's suicide belt." She has repeatedly accused Monsanto of committing "genocide" in India because farmers cannot afford to plant the Bt cotton that the company introduced and thus have accumulated burdensome debt. As late as 2014, she claimed, "Farmers are dying because Monsanto is making profits by owning life that it never created but it pretends to create. . . . That is why we need to get rid of the GMOs."[89] The number of suicides in Maharashtra increased from 200,000 in 2009 to 284,000 in 2014,[90] but the charge has not been upheld under international or Indian scrutiny.[91] The American political scientist Ronald J. Herring, for example, identified a seeming paradox: "It is hard to imagine farmers spreading a technology that is literally killing them."[92]

Shiva's accusation that Bt-cotton seeds have been created almost exclusively to serve large industrial farms is partially true. At first, such seeds were extremely expensive in India, as elsewhere, leading to counterfeits that flooded the market. The fake crops failed, and many farmers suffered. Consequently, according to Shiva, Bt-cotton seeds had cost eighty times more in India since 2002; but in reality, with government's regulation, the price of Bt-cotton seeds had fallen steadily. While their price remains higher than that of conventional seeds, in most cases the Bt seeds provide greater benefits. According to the International Food Policy Research Institute at Stanford University, while Bt-cotton farmers spend at least 15 percent more on crops, their pesticide costs are 50 percent lower because of significant reductions in pesticide usage, and yields have increased by more than 150 percent. In 2012, India's agriculture minister acknowledged that the country "has harvested an average of 5.1 million tonnes of cotton per year, which is well above the highest production of 3 million tonnes before the introduction of Bt cotton."[93]

However, Shiva's claim that Monsanto's patents had prevented poor cotton farmers from saving seeds is not true. In fact, the Farmers' Rights Act of 2001 guarantee every Indian farmer the right to "save, use, sow, resow, exchange, share, or sell" seeds. To ensure better yields and bigger profits, most farmers, even those whose plot is as small as the backyard of a house in the American suburbs, have chosen to buy newly bred seeds each year, whether genetically modified or not.[94]

Summary and Discussion

The United States and the European Union represent polar opposites in the heated global debates on GM crops and foods. Indeed, they have taken very different policy approaches to this same technology. The United States adopted GM crops and foods much earlier, framed the development of GM crops mainly from an economic-competitiveness perspective, and formulated its policy under a more liberal environment. Moreover, the United States has not experienced serious food-safety scandals; Americans are generally more optimistic about technological breakthroughs and in favor of innovation; and they have a more trusting perception of their regulators and scientists, at least at the time when the GMO policy was introduced. While recognizing GM crops as substantially equivalent to their conventional counterparts, U.S. regulators have been precautionary in dealing with environmental issues associated with GMOs but have not taken the same steps to safeguard against GM foods as their European counterparts have. With a relatively smoother ride, the United States has also witnessed GMO-related environmental and food-safety controversies, including the "monarch butterfly" research; the recall of food products contaminated with StarLink corn, a GM variety approved for use as animal feeds only; and contamination of the American grain supply by LibertyLink, a GM rice that had not been deregulated.

On the other side of the Atlantic, out of food-safety concerns, the EU has been preventive in its policy toward GM crops and especially GM foods, although the European Commission also agrees that approved GMOs are safe for health and the environment.[95] Relatively new for the public, GM crops and foods are more controversial in the EU than in the United States, and even ideologically and politically charged, so that a large European population has been somewhat antipathetic and even hostile to the application of biotechnology in agriculture as opposed to medicine.[96] Defining each GM food as "novel," European regulators require extensive premarket risk assessment. The EU has introduced additional and often tougher measures on contained use, deliberate release, and labeling. With a moratorium in place since 1999, the number of GM products available has been significantly limited.

In light of the introduction of CRISPR/Cas9, the new gene-editing technology, the United States and the EU may converge on their relevant regulations. For example, in 2016, USDA decided not to regulate the cultivation and sale of a CRISPR-created white-button mushroom, which, with specific genes deactivated, is less likely to turn brown or bruise and so is more resilient to automated harvesting and has a longer shelf life. The approval was not unexpected given the "substantial equivalence" principle.[97] Indeed, on March 28, 2018, USDA announced to do the same on other types of gene-edited crops.[98] Meanwhile, European scientists urge the EC not to stifle the powerful gene-editing tool by lumping it together with transgenic technology.[99] While debates have been going on in Sweden and Germany regarding regulation of CRISPR-edited plants, early indications seem to suggest that gene-edited organisms and foods may still have an uncertain future in the EU.[100]

Policy on GM crops and foods in India has evolved within this extremely divergent international environment. Apparently, the "elaborate" division and sharing of responsibility by several government agencies was modeled on the American practice. The rising anti-GMO activism and its adverse impact on the trajectory of transgenic technology have had clear European influence, making India's regulatory regime "rigid." The presence of Vandana Shiva only shows that activists can be more radical than their European counterparts, whose most famous representative, Mark Lynas, recently reversed his anti-GMO position.

Against the background of this global policy environment toward transgenic technology and GMOs, it is inevitable that various stakeholders in China, from policymakers, scientists, and NGOs to consumers, have drawn on experiences and learned lessons from their international counterparts.

CHAPTER III

Research and Commercialization of GM Crops in China

Biotechnology features prominently in China's knowledge-based economic development strategy and its bid to become an innovation-oriented nation. In particular, China has placed big bets on agricultural biotechnology and GM crops, hoping not only to ensure food security—meeting its increasing demand for food with higher yields, improved nutrition, and better resistance to pests and diseases while alleviating the huge problem of pollution from excess pesticide use—but also, more importantly, to become a global leader in biotechnology.[1] With the state playing virtually every role in biotechnology's development, an enthusiastic scientific community has made great efforts to achieve remarkable progress in the past three decades.

Chinese farmers started to grow the world's first GM crop, a virus-resistant tobacco, albeit illegally, in 1988.[2] In 1992, cultivation of GM tobacco was approved, making China the first GM nation. The hectarage devoted to GM tobacco reached one million in 1996 and increased to 1.6 million in 1997, which was then the world's largest GM cultivation. By 1996, some forty-seven crops had become transgenic in labs or in experiment fields, including five grain crops, five cash crops, ten vegetables, eleven fruit trees, three forest trees, two fodder crops, and five medicinal and ornamental plants.[3] In 1997, China approved the commercialization of GM crops of Bt cotton and slow-ripening tomatoes.[4]

However, research on and commercialization of GM crops are not confined to scientific research, or the upper stream of the development of GMOs. As I will argue, China's agricultural biotechnology program is situated in a policy environment that has been generally promotional but fast changing and evolving, not unaffected by "other" factors. This chapter will also examine the role played by foreign technology embodied in the training of Chinese biotechnological talent at the international frontier of transgenic research. The chapter ends with a comprehensive analysis of the development of Bt cotton in China.

S&T Policymaking in China: An Overview

It is well known that China has a sophisticated bureaucratic system for making policies related to its various objectives.[5] At the front of science and technology, policy is formulated, as elsewhere in the world, by the interaction of scientific and political institutions involving actors from legislature, government, advisory bodies, performing organizations, and funding agencies.[6] In particular, the National People's Congress (NPC), China's highest organ of state power and legislature, at least on paper, through its Standing Committee and the Committee on Education, Science and Technology, Culture, and Health, has the authority to enact and amend all S&T-related laws, which are mostly drafted by a ministry or several ministries under the State Council, China's cabinet. NPC also monitors the implementation of such laws and approves the state budget on science and technology.[7] During the policymaking process, members of the Chinese People's Political Consultative Conference (CPPCC), an advisory body with many non-Communist-Party-member scientists, also have a voice. An example of the active role played by CPPCC members in policymaking was their participation in the debate on the Three Gorges dam project across the Yangtze River in the 1980s. Concerned with the possible social, economic, and environmental impacts of the project, scientist members of CPPCC urged that the project be reevaluated. In 1983, led by Zhou Peiyuan, a physicist and CPPCC vice chairman, a group of scientists and engineers undertook a major feasibility study and concluded that such a project was not feasible scientifically, economically, and environmentally. They succeeded, at least temporarily, in delaying the project and forcing the government to revise it.[8]

Ministries under the State Council with missions in S&T formulate rules, regulations, and other policy measures related to their respective institutional mandates and the nation's development strategy for the purpose of implementing laws. The National Development and Reform Commission (NDRC) is probably the most important agency in the formulation and implementation of policies pertaining to S&T. The Ministry of Science and Technology (MOST) oversees the nation's S&T enterprise by exercising a wide range of functions, from formulating S&T policies, plans, and programs to allocating S&T budget and resources. The ministries of education (MOE), agriculture and rural affairs (MOARA), ecology and environment (MOEE), and industry and information technology (MOIIT), and the National Health Commission (NHC) are responsible for making policy relevant to their respective missions.[9] The Ministry of Finance (MOF) has become increasingly important in scrutinizing budgets put forward by government agencies and monitoring the spending of the funds.[10] As will be discussed in the next chapter, many of these ministries are also the main players in policymaking on agricultural GMOs. Given the fragmented nature of China's bureaucracy, such policy is often arrived at through bureaucratic negotiations and bargaining.[11]

Nonetheless, it is the Chinese Communist Party (CCP) that has the final say in S&T policymaking. While not enacting laws directly, CCP exercises its influence and power in policymaking through a variety of indirect means that reflect its continued clout in virtually all matters. CCP's Central Committee (CCPCC) not only sets S&T policy but also leads science, technology, and education through a State Leading Group for Science, Technology, and Education within the State Council. Normally, a leading group is established as a mechanism through which the Chinese central government tackles issues and coordinates efforts involving multiple ministries. Given the importance attached to science, technology, and education in the reform and open-door era, a State Leading Group for Science and Technology was set up in 1983 and renamed in 1998. Comprised of heads of all ministries involved in China's S&T enterprise, the leading group is chaired by the premier, who is concurrently a member of CCPCC's Politburo Standing Committee—China's de facto top governing body. It is charged with studying and reviewing the nation's strategy and key policies for the development of science, technology, and education; for discussing and reviewing major tasks and programs related to science, technology, and education; and for coordinating important issues of science and education involving agencies under the State Council and regions.[12] The State Leading Group for Science and Technology Structural Reform and Innovation System Construction, established in September 2012, takes the lead in comprehensive planning and coordination; it deliberates major tasks, major policies, major questions, and major work arrangements regarding reform of the S&T system.[13] The new Central Commission on Comprehensive Reform and the new Central Commission on Financial and Economic Affairs, both party apparatus upgrading the Central Leading Group for Comprehensive Reform and the Central Leading Group for Financial and Economic Affairs respectively in 2018, seem to have taken over or at least integrated many responsibilities that used to be mandated for the leading groups under the State Council. CCPCC also reviews all major initiatives before sending them to NPC for legislation or to the State Council and its ministries for specification and implementation.[14] In general, a high-level policy guides the macro and strategic direction, while a policy issued by a low-level government agency is largely focused on the specification and interpretation of the high-level policy, the operational mechanism associated with the macro policy, or agency-specific measures.[15]

Compared to the fairly clear policymaking structure and division of labor between the party and the state and between ministries, the process through which a particular policy is proposed, deliberated, and formulated in China remains a "black box." Policymaking starts with agenda-setting. In China, policy agenda-setting is either top-down or bottom-up, involving stakeholders such as policymakers themselves, think tanks, interest groups, NGOs, and the public. Specifically, there are six ways to set a policy agenda, depending upon the level of public participation.[16] Indeed, public participation has become especially critical since 1986, when Wan Li, then a member of the Politburo and Secretariat of CCPCC and a vice premier, called for scientification and democratization in policymaking.[17] Ultimately, policymaking ought to balance the interests of all stakeholders. The scientification goal, if policymaking has nothing to do with the distribution of interests, may be pursued mainly through engaging with the elite; democratization involves public participation, so that the public, broadly defined, is involved in policy agenda-setting.

Agendas of S&T policy are set through closed-door discussions or open-door petitions by individual scientists, institutions, and organizations.[18] The issuance of a central document, to be discussed, falls into the closed-door model in which CCPCC, often together with the State Council, sets a policy agenda. Petitioning individually, collectively, or institutionally to set an agenda is exemplified by the initiation of the State High-Tech Research and Development Program, better known as the 863 Program,[19] and the Knowledge Innovation Program of the Chinese Academy of Sciences (CAS).[20] The open-door model of S&T policymaking mobilizes the participation of the scientific community, especially its elite members, think tanks, and other stakeholders. The formulation of the Medium- and Long-Term Plan for the Development of Science and Technology (2006–2020) (MLP), issued in 2006, as a State Council document, is a case in which both models were used. While most deliberations were conducted behind closed doors and not made public, more than 2,000 scientists, engineers, and corporate executives participated in an early phase of "strategic research" to identify critical problems and research opportunities in twenty areas considered to be of central importance to China's future. Foreign experts were also consulted. It was only at a later stage that the openness of the "strategic research" phase gave way to a closed-door process during which bureaucrats massaged the strategic research group reports and attempted to reach compromises in what was by most accounts a contentious and unusually drawn-out process of drafting the public version of the plan. Narrowing MLP's focus as well as setting research priorities proved to be so onerous that at one point it required direct and top-down intervention by Premier Wen Jiabao![21] Of course, agenda-setting is not a clear-cut process as even the drafting of a central document in its stage of conceptualization is likely to involve engaging or at least consulting various stakeholders with interests in the document.

Scientists are active in policymaking related to S&T development in general and to their disciplines in particular. In this regard, elite members (*yuanshi*) of the CAS and the Chinese Academy of Engineering (CAE) are expected to fulfill their collective role of making these two academies the nation's top "brain banks," advising the state with their expertise. In fact, the elite institution bestows on Chinese *yuanshi* the right and responsibility to render their independent scientific judgments and consultation. CAS and CAE members have been instrumental in outlining and evaluating the nation's major science programs in the reform and open-door era, and if they want, they can—and did—submit suggestions to CCPCC's Politburo.[22] The governing political elite, until most recently, were often trained at the country's top science and engineering schools and were naturally inclined to favor policies that promoted scientific advances.[23] As I will illustrate, while central documents represent the leadership's political will, petitions to the leadership have become a critical avenue for Chinese agricultural biotechnologists, especially the elite members, to set the policy agenda and voice their concerns to policymakers on the development of transgenic technology.

Central Documents No. 1

Usually listing focal points for work in a particular area, documents are one of the means by which the Chinese party-state makes policy and governs the nation. Not all the documents are of equal weight. A central document (*zhongyang wenjian*) issued by CCPCC, often together with the State Council, is most authoritative and supposedly most influential and effective, usually overriding a law enacted by NPC, an administrative statute formulated by the State Council, or a regulation issued by ministries.[24] In other words, regardless of the content of a specific policy, the political weight of a central document is greater than that of a law, which, in turn, is greater than that of a State Council administrative statute or a ministerial regulation.

Moreover, the first central document issued in a year (*zhongyang yihao wenjian*) usually sets the priority of the party and the central government for the year, symbolically as well as substantively. In a significant number of years in the reform and open-door era—between 1982 and 1986 and between 2004 and 2016—Central Documents No. 1 issued by CCPCC and the State Council have been about agriculture, rural affairs, and farmers (*sannong wenti*). This focus is not surprising given that agriculture still contributes significantly to the Chinese economy, and the rural population, albeit declining, still accounts for 42 percent of the total. What is most extraordinary is that in recent years, Central Documents No. 1 have repeatedly invoked transgenic technology and GMOs, with varied tones (table 3.1).

The GMO issue was first brought up in 2007's Central Document No. 1, which was about strict implementation of labeling GM foods. In the following year, Central Document No. 1 embraced transgenic technology by calling for launching the Mega-Engineering Program (MEP) for the cultivation of new varieties of GMOs, one of MLP's MEPs. In 2009, in addition to stressing the acceleration of MEP, Central Document No. 1 called on integrating research resources; increasing investment in R&D; cultivating a number of new GM crop varieties with characteristics of disease and insect

TABLE 3.1

Transgenic Technology and GMOs in China's Central Document No. 1

Year	Important Development
2007	Strictly implement the labeling of GM foods.
2008	Launch the Mega-Engineering Program (MEP) for the cultivation of new varieties of GMOs.
2009	Accelerate the MEP for the cultivation of new varieties of GMOs, integrate research resources, increase investment in R&D, cultivate a number of new varieties of GM crops with characteristics of disease and insect resistance, stress tolerance, higher yield, higher quality, and higher efficiency as soon as possible, and promote commercialization.
2010	Continue to implement the MEP for the cultivation of new GMO varieties, accelerate the development of functional genomics and new biological varieties with important application potential and indigenous IPRs, and promote the commercialization of new GMO varieties based on scientific assessment and management as well as relevant laws and regulations.
2012	Achieve major breakthroughs in basic theory and methodology of gene regulation in agricultural biotechnology, molecular breeding (*fenzi yuzhong*), and biosafety. Continue to implement the MEP for the cultivation of GMO new varieties.
2014	Strengthen basic research and biotechnology development, focusing on molecular breeding.
2015	Strengthen research, safety management, and scientific popularization of agricultural transgenic biotechnology.
2016	Strengthen R&D and supervision of agricultural transgenic technology and promote the technology prudently on the basis of ensuring safety.

resistance, stress tolerance, higher yield, higher quality, and higher efficiency as soon as possible; and, most importantly, promoting commercialization to benefit Chinese farmers. The year 2010 was the first time that Central Document No. 1 promoted the development of GMOs not only as a way of improving crop-breeding but also as a national strategy for agricultural development. In particular, the document stipulated continuing implementation of MEP, accelerating development of functional genomics and new biological varieties with important application potential and indigenous intellectual property rights (IPRs), and promoting the commercialization of new GMO varieties based on scientific assessment and management as well as relevant laws and regulations.

Apparently, the GMO MEP was favored not just by the transgenic research community but also by the central government, which by 2010 had referred to the program in its Central Document No. 1 three years in a row. The message seemed loud and clear, and it was followed by significant and concrete moves. In July 2008, the GMO MEP was formally launched; it would channel some RMB20 billion ($2.9 billion), more than half from the central government and the rest from local governments and industry, by the end of the plan period in 2020.[25] In August 2009, the Ministry of Agriculture (MOA) issued biosafety certificates for commercialization of two strains of Bt rice and phytase maize.

While acknowledging that Central Document No. 1 of 2010 marked a new phase for the research and commercialization of transgenic technology, Deng Xiuxin, president of Huazhong Agricultural University, which had two strains of Bt rice recently granted biosafety certificates, still interpreted the central government's attitude toward the technology as "active but steady." Indeed, the government is "active" in promoting the acceleration of research on and commercialization of GM crops but remains "steady" or "precautionary" in its emphasis on stringency regarding biosafety regulations.[26] There is nothing wrong with subjecting the development of GM crops to a rigid biosafety regulatory regime, but the issuance of biosafety certificates to GM varieties soon stirred considerable opposition from various quarters, which will be discussed in the following chapters.

The next time that Central Document No. 1 brought up the topic of transgenic technology was 2012. It would have been remarkable had the document, which was about science, technology, and innovation in agriculture, said nothing about transgenic technology. Along with the reference to the strengthening of basic research so as to achieve major breakthroughs in agricultural biotechnology, especially in such areas as genetic regulation and molecular breeding, the document again underlined GMO MEP's continuous implementation. It is also worth noting that "molecular breeding" (*fenzi yuzhong*), a more subtle but more accurate term than transgenic technology, was introduced,[27] which some scientists regarded as referring to the same technology while others disagreed.[28]

In 2013, Central Document No. 1 said nothing about GM crops, MEP, or molecular breeding. Instead, it alluded only vaguely to continuous implementation of key S&T programs such as breeding development. The next year's Central Document No. 1 mentioned strengthening basic research and biotechnology development focusing on molecular breeding, without mentioning the GMO MEP and the acceleration of R&D on GMOs, let alone commercialization.[29] This represented a sea change regarding GM crops, which was not entirely unexpected in light of the campaigns that activists had mounted against commercialization of if not research on GM crops, especially GM rice. In 2015, Central Document No. 1 emphasized research, safety management, and popularization of agricultural transgenic biotechnology. This document for the first time also discussed making efforts to communicate to the public about GMOs to increase awareness and then acceptance of GM crops, obviously out of the concern that the GMO issue had not previously been properly conveyed to the public. In 2016, Central Document No. 1 called for strengthening R&D, monitoring agricultural GMOs, and promoting GMOs with caution regarding safety.

Between 2007 and 2016, these themes—agricultural genetic engineering, transgenic technology, and the development of GM crops—have appeared eight times, albeit very briefly, in *sannong*-oriented Central Documents No. 1. The tones, representing the attitudes of China's political leadership toward research on and commercialization of GM crops, have alternated between promotion and precaution, as have the views of different stakeholders involved or influential in the drafting of the documents. Most recently, the years 2012 and 2016 witnessed a back-and-forth shift in the policy landscape of transgenic technology within an overall uncertain domestic environment. The central documents have emphasized the strengthening of basic research and biotechnology development centered on molecular breeding, not transgenic technology; the strengthening of safety management as well as popularization of transgenic technology; and the promotion of cautious development of GMOs. While not overemphasizing or exaggerating the exact role of Central Documents No. 1 in guiding the development of transgenic technology, I would argue that these documents have led to concrete actions to substantiate the leadership's supportive attitudes toward the issue. In particular, these important political manifestations have made it easier for the research community to advance its cause.

The Central Leading Group on Rural Work, a party apparatus, often holds a news conference to interpret and elaborate on the issues raised in a *sannong*-related central document, and questions relating the application of transgenic technology in agriculture and GM crops have come up time and again. For example, in 2014, Chen Xiwen, a deputy director of the leading group, put the party's positions on GMOs into perspective. First, as a large agricultural country, China cannot afford to lag behind in molecular breeding, a cutting-edge technology of the life sciences, and should strive to catch up with the leading technology in the world. Second, regarding the sale of GM crops and foods and making them available to consumers, China must undergo an extremely rigorous assessment and only approve and commercialize those that have no adverse effects. Third, China must offer consumers a full "right to know" (*zhiqingquan*), a term to which I will return, and it is up to consumers whether to buy GM products. "We will adhere to these basic principles for a long time," Chen explained, "and we also will not change the two long-term principles" of conducting rigorous safety assessment and inspection and providing consumers with the right to know.[30]

During the same press conference, Chen Xiwen also discussed the global attention to safety issues associated with GM varieties created through molecular breeding. Transgenic breeding was first developed in the United States, said Chen, and GM foods have been put on the market only after the U.S. Food and Drug Administration approved them safe for human consumption. While there have been debates, the United States has generally accepted GM foods, required no labeling, unlike other countries including China. Chen appeared to imply the American influence on China's policy on one hand, and China's ambivalence toward its policy, which had led the government to retreat from its previous promotional position, on the other.

At the press conference after the release of 2015's first central document, Han Jun, also a deputy director of the Central Leading Group on Rural Work, emphasized that China's GMO policy has been consistent—that is, to strengthen research on agricultural biotechnology and GMOs' biosafety simultaneously. Transgenic technology represents the future development of new technologies and new industries. Although there still is a significant gap between China and advanced countries in the technology, China began research on GMOs early on, thus leading in some areas, especially on GM rice and corn. Facing environmental and resource constraints in feeding a population of 1.3 billion, China cannot afford to fall behind in the study of transgenic biotechnology. However, according to Han, the central document also noted the importance of enhancing the safety management of GMOs. China's biosafety regulations have been established by drawing not only on its own national conditions but also on international experiences and standards, covering research, testing, production, processing, imports, and labeling in a comprehensive way. The government has the responsibility to implement relevant laws and regulations and take on activities that violate them.

Han Jun pointed out that transgenesis should be recognized as a scientific issue, which might not be quite comprehended by those who are not involved in the research. The GM issue also becomes very sensitive concerning the entire society and relating to the daily life of ordinary citizens. Therefore, disseminating relevant and accurate knowledge is necessary to help the public and the media to achieve a comprehensive, objective, and thorough understanding of how the technology has developed, what the characteristics and risks of the technology are, and how China has established and improved its safety management system against those of other countries. China needs to respect science and be more rational toward transgenic technology and GM products, Han lamented.[31]

At the 2016 news conference, Han Jun again expressed his deep understanding of public concerns about GMOs and affirmed the government's assurance regarding consumers' right to know and to choose, on top of regulations and oversight, including mandatory labeling of GM foods. In answering the question of whether the document's careful promotion of GM crops on the basis of ensuring safety represents a change of policy, Han again indicated that China's national policy on transgenic technology has been consistent and unchanged. He noted the coming of age of the new S&T revolution in agriculture in which China had been left behind. Han also said that China needs to tackle the illicit cultivation of GM rice and strengthen supervision and enforcement of the regulations, particularly focusing on R&D, variety approval, and production and distribution, to prevent seeds of illegal GM crops from entering the market. Meanwhile, Han Jun indicated that careful promotion means undergoing a rigorous validation and approval process. Commercial planting of GM crops has been expanding globally amid some controversies, but the international consensus on the safety of GM products is authoritative—that approved GM foods are as safe as conventional ones and that all the commercially cultivated GMOs have undergone not only research but also a rigorous safety assessment.[32]

At all these news conferences, both Chen Xiwen and Han Jun also mentioned that China should occupy the high ground of transgenic technology and that China's GMO market should not be saturated with foreign products, which was fairly nationalistic. It turns out that they just repeated what Xi Jinping, general secretary of CCPCC and China's state president, had said. In his speech at the conference of the central rural work held on December 23, 2013, just days before the release of 2014's Central Document No. 1, Xi discussed the transgenic issue. He said that transgenic technology is a new technology, representing a new industry with a broad prospect for development. Because of the novelty, it also is quite normal for the public to have doubts about it and debate the technology. Xi then emphasized safety and indigenous innovation:

> Industrialization and commercialization of transgenic crops must strictly follow technical regulations and specifications formulated by the state, proceeding steadily to ensure no mishap and taking into account all safety related factors. We should boldly carry out research and innovation, take the high ground of transgenic technology, and do not allow foreign companies to saturate our agricultural GM product market.[33]

These remarks represented a balanced but slightly promotion-leaning position on transgenic technology and GMOs. Such a policy development coincided with the change of political leadership in China in late 2012 and early 2013 when Xi Jinping and Li Keqiang took over the party and state administration respectively. However, GMOs were not new to the new leadership. In retrospect, on September 12, 2012, Xi Jinping, then a member of CCPCC's Politburo Standing Committee and state vice president, learned firsthand about Bt and drought-resistant corn during a visit to China Agricultural University.[34] One year later, on September 30, 2013, Xi and other members of CCPCC's Politburo were briefed on the Chinese progress in GMOs during a group study at Beijing's Zhongguancun Science Park, although then neither Xi nor other members of the political leadership commented on the issue publicly.[35] Therefore, the change of policy may also reflect the fact that the new leadership needed some time to decide on its preferences and priorities in agricultural technology, which was not unexpected by the scientific community.[36] Seen from this perspective, I can only conclude that China's political leadership has now recognized the national interests in the development of indigenous transgenic technology.

Policy Environment for Development of the Life Sciences/Biotechnology

China's reform and open-door era has witnessed the central government's promotion of the development of science and technology. In the immediate aftermath of the Cultural Revolution, the 1978 National Science Conference renewed Chinese enthusiasm about S&T modernization. An eight-year (1978–1985) scientific development plan prioritized eight areas of research, including agriculture and genetic engineering, as well as energy, new materials, computers, lasers, space, and high-energy physics (table 3.2). These areas were given further notice when China launched the 863 Program in 1986. Amid the worldwide new technology revolution, four senior scientists who had been involved in China's strategic weapons program—Wang Daheng, Wang Ganchang, Yang Jiachi, and Chen Fangyun—wrote to Deng Xiaoping, then China's paramount leader, suggesting following the world's high-tech trends and developing high technology in China. Deng and other political leaders immediately endorsed the petition, and the 863 Program was soon outlined under both the State Science and Technology Committee (SSTC), the predecessor of MOST, and the Commission of Science, Technology, and Industry of National Defense, with RMB10 billion ($2.9 billion) to be allocated over fifteen years. The program's priorities were similar to those of the eight-year plan, placing significant emphasis on seven technological areas (*lingyu*), including again biotechnology (the other six areas were space, information, lasers, automation, energy, and new materials, with marine technology added in 1996).[37] The 863 Program was the first instance in which a group of Chinese scientists set an S&T policy agenda by expressing their collective interests to the political leadership. Of course, deliberation of the priorities mobilized the participation of a large community of scientists and other stakeholders in high-tech development.

The inclusion of biotechnology in the 863 Program also saw some influence from Chinese-American scientists. The United States already had a tremendous lead in information technology, but biotechnology was just

TABLE 3.2

China's Biotechnology-related Research and Development Programs

Name	Year Introduced	Government Agency in Charge	Key Development
Key Breakthrough Science and Technology Program	1982	SDPC	Updated every five years. One of the major components of these projects is biotechnology R&D.
National Key Laboratories (NKLs) Program	1984	SDPC	Thirty NKLs in biotechnology (fifteen in agriculture or agriculture-related) have been established.
National Biotechnology Development Policy Outline	1985	SSTC, SDPC, and others	The document defined the research priorities, development plan, and measures of achieving targets.
State High–Tech Research and Development Program (863 Program)	1986	SSTC	Initially with RMB10 billion for fifteen years to promote high–tech R&D. Biotechnology was one of the original seven priorities with a total budget of about RMB1.5 billion between 1986 and 2000.
National Natural Science Foundation of China (NSFC)	1986	NSFC	Support basic research and mission–oriented basic research (*yingyong jichu yanjiu*).

Program	Year	Agency	Description
State Basic Research and Development Program (973 Program)	1997	MOST	Support basic science and technology research in six areas, including population and health, information, agriculture, resources and the environment, energy, and new materials.
High-Tech Commercialization Special Program	1998	SDPC	Promote the application and commercialization of high technology.
GM Plant Research and Commercialization Special Program	1999	MOST	A five-year program to promote the research and commercialization of transgenic plants.
Medium- and Long-Term Plan for the Development of Science and Technology (MLP)	2006	State Council	The cultivation of new varieties of GMOs is one of the sixteen Mega-Engineering Programs.

Note: The names of the government agencies are the ones used when the programs were initiated: SDPC = State Development and Planning Commission, now part of the National Development and Reform Commission (NDRC); SSTC = State Science and Technology Commission, predecessor of the Ministry of Science and Technology (MOST).

getting off the ground, they told Chinese leaders; therefore, China should have an easier time catching up and even becoming competitive. On January 10, 1983, for example, Ray Wu, a Chinese-American professor of molecular biology at Cornell University, wrote to Fang Yi, a State Councilor in charge of science, emphasizing the benefits of and opportunities for investing in education and research in the life sciences.[38] On September 13, 1986, just two days after meeting Chinese-American physicist Chien-Shiung Wu, who suggested using bioengineering to solve China's agriculture problem, Deng Xiaoping remarked, "the ultimate solution to agriculture in the future lies in bioengineering and sophisticated technology."[39]

Of the fifteen subjects (*zhuti*) designated within the initial seven areas of the 863 Program, two were related to agricultural biotechnology: genetic engineering for high-yield, high-quality plants and animals, and protein engineering for food industries and agriculture. The special projects (*zhuanxiang*) supported by the 863 Program in its first fifteen years also included the rice genome. In general, the 863 Program followed the international lead to target mainly applications in agriculture as well as pharmaceuticals (green and red biotechnologies).[40] It also set aside funds specifically to aid commercialization of research outcomes, the weakest link in China's research enterprise.[41] The program began China's efforts to research on and then commercialize agricultural GMOs. In 1988, SSTC launched a Torch Program (*huoju jihua*) to bring research closer to the economy, to create markets for high technology, to commercialize inventions, and to promote academic entrepreneurship.

Despite having the basic research components underlying high technology, the 863 Program was in essence a program of following and catching up. Thus, in 1989, SSTC initiated a Climbing Program (*pandeng jihua*) to make up for the neglect in basic science and mission-oriented fundamental research. The program sought to attract China's brightest scientists nationwide to work on thirty critical research areas, from high-temperature superconductivity to nonlinear science to important chemical problems in life process to brain function and its cell and molecular bases. The program evolved into a State Basic Research and Development Program, or the 973 Program, in 1997 at the suggestion of scientist delegates to NPC and scientist members of CPPCC.[42] The mission-oriented 973 Program supported research projects falling within six broad areas relevant to the nation's economic and social development: population and health, information, agriculture, resources and the environment, energy, and new materials. In order to showcase China's endeavors in pursuing excellence in basic research, projects included in the 973 Program were supposed to meet one of three criteria. First, the projects attempted to solve major problems associated with China's social, economic, and technological development. Second, they were related to major basic research problems with interdisciplinary and comprehensive significance. Third, they were able to leverage China's advantages in natural, geographic, and human resources and help China occupy "a seat" (*yixi zhidi*) at the league table of international research.[43]

The mission orientation in scientific research became most salient when the state formulated its fifteen-year S&T MLP in the early twenty-first century. As a roadmap for China's ambition to become an innovation-oriented nation by 2020 and a world leader in S&T by 2050, the plan selected the life sciences and biotechnology as priorities that were expected to make a significant difference. Biotechnology was again one of the eight areas of "frontier technology"; GMO new varieties, along with drug innovation and development, AIDS, viral hepatitis, and the control and treatment of other major diseases, were among the sixteen MEPs; and protein science and developmental and reproductive biology were included in the Mega-Science Programs (MSPs).[44] The government later identified biotechnology as one of the seven strategic emerging industries and included it in the Made in China 2025 initiative and the Thirteenth National Five-Year Plan for Science, Technology, and Innovation (2016–2020) to provide solutions to China's economic restructuring, underpinning long-term economic growth and serving as a pathway for building indigenous innovation (*zizhu chuangxin*) capability.

The development of GMOs in agriculture has thus been situated in a largely supportive policy environment for the development of the life sciences/biotechnology. The nurturing of such an environment has also benefited from, or is a result of, the interaction between the scientific and political leadership. The agricultural biotechnology research community has seen and seized on opportunities for political support by actively propagating their views and advancing their interests. Given its technocratic orientation, China's political leadership has frequently invited leading academics to update its members on various aspects related to the nation's social and economic development, including scientific knowledge. In 2001, for example, the Leading Group for Science, Technology, and Education invited Zhang Qifa, one of the leading plant geneticists at Huazhong Agricultural University and a CAS member (*yuanshi*), to lecture on genetic engineering and

GM foods. On that occasion, Premier Zhu Rongji expressed support for increasing investment in S&T and human resources and for promoting the commercial planting of GM crops. In 2004, the CAS Academic Divisions (*xuebu*), comprised of elite CAS members, initiated a consulting project on the development of GM crops. Zhang Qifa led a group of sixteen scientists, economists, and food-safety experts involved in GM-rice research to petition the central government to loosen some of the regulatory controls over GM crops and accelerate the commercialization approval process. They noted that despite billions spent on research, the nation had yet to approve a single GM crop for commercialization after 1997, when Bt cotton was given the green light, condemning an "unclear policy direction." They said that overly restrictive regulations and lengthy approval processes "suffocated the development of transgenic technology in China."[45] The project, along with other similar efforts, eventually led to the inclusion in MLP in 2006 of a mega-engineering program for GMO new varieties, which was subsequently launched in 2008.[46] On June 25, 2010, NPC's Standing Committee invited Huang Dafang, a leading agricultural biotechnologist from the Institute of Biotechnology of the Chinese Academy of Agricultural Sciences (CAAS), to lecture on the technology to its members.[47] Most recently, as mentioned, GM scientists were among those who briefed the Politburo of CCPCC when it held its group study at Zhongguancun Science Park.

In the countdown to the expiration of biosafety certificates issued to Bt rice and phytase maize, three groups of elite scientists involved in the research on and commercialization of GM crops petitioned the political leadership. On November 30, 2012, at a forum organized by CAE, fifty CAS and CAE *yuanshi*, led by former CAS vice president Shi Changxu, signed a petition calling for the acceleration of commercialization of the GM-crop research achievements; the petition was submitted to the central leadership in early 2013. In June 2013, thirty-two CAE *yuanshi* of agriculture, led by Dai Jingduan, a GM-corn researcher, petitioned the State Council to speed up research on GMOs and commercialization of GM staple crops. Then in July, Zhang Qifa led sixty-one CAS and CAE *yuanshi* working on GMOs to petition the central leadership again. In addition to advocating the acceleration of commercialization, scientists requested immediate authorization to plant Bt rice, putting pressure on the leadership by particularly singling out MOA for inaction.[48]

Such frequent and high-level access has generated a gradual but significant influence on the central leadership, which has responded with some high-profile statements. In his interview with Bruce Albert, editor-in-chief of *Science*, on September 30, 2008, for example, Premier Wen Jiabao "strongly advocated making great efforts to pursue transgenic engineering." The premier also evoked the influence of recent global food shortages that had further strengthened his belief in developing transgenic technology.[49] In his report to the third session of the Eleventh NPC in March 2010, Premier Wen further pointed out that "with the focus on breeding improved crop varieties, we will accelerate innovation in agricultural technology and the widespread adoption of scientific and technological advances, and carry out a mega-engineering program to create new crop varieties using transgenic technology."[50] Such remarks, followed by the launch of the GMO MEP and the biosafety certification of Bt rice and phytase maize, corroborate that the promotional tone of several Central Documents No. 1 did represent the political leadership's stance on GM crops. However, the petitions in 2013 seem not to have garnered sufficient leadership attention, as biosafety certificates for GM crops still were allowed to expire in August 2014.

China's Life Sciences/Biotechnology in Historical and Global Context

Perceived to have significant global impact in decades to come, the life sciences and biotechnology have gained prominence among emerging S&T fields in China. However, the nation's interests in them date back quite some time. In the Republican period (1911–1949), scientists focused on and made achievements in applied scientific fields related to China's national conditions, including biology and agriculture as well as geology, mining, paleoanthropology, paleontology, geography, and meteorology. Meanwhile, biochemistry stood out, along with mathematics, physics, and chemistry, in basic science.[51]

After the founding of the People's Republic in 1949, scientific research was interrupted by political upheavals, especially the Anti-Rightist Campaign in 1957 and the Cultural Revolution between 1966 and 1976. However, some Chinese scientists still sought opportunities to practice and pursue their professions and contribute their knowledge and expertise to nation-building. The life sciences were part of China's overall S&T efforts early on, as stipulated in the twelve-year (1956–1967) S&T program. Although it was geared toward the development of strategic weapons—atomic bombs, missiles, and satellites (*liangdan yixing*)—the program did not completely ignore theoretical and fundamental research in biology, as well as mathematics, mechanics, astronomy, physics, geology, and geography that underlay weapon projects and could have enormous societal impacts. A glance at some of the country's life sciences accomplishments back then reveals the efforts toward discoveries with high societal relevance. In the early 1960s, for example, Chinese life scientists successfully synthesized bovine insulin, which, as the "world's first," was believed to be a major achievement of the caliber of a Nobel Prize. Amazingly, research on x-ray crystallography of insulin, study of the structure and activity of insulin analogy, and synthesis of other polypeptide hormones, among other investigations, were allowed to continue during the Cultural Revolution when almost all scientific research activities were severely stalled.[52] In 2015, Tu Youyou received China's first science Nobel Prize—in Physiology or Medicine—for the use of artemisinin in the treatment of malaria, a discovery that also was achieved during the Cultural Revolution.[53] Of course, had there not been severe disruptions targeting particularly scientists and other professionals, China could have made more progress in science and technology.

After the Cultural Revolution, China resumed its attention to science and technology, regarding them as critical for modernization in agriculture, industry, and national defense. The S&T system was among the first to be reformed. More enlightened and sustained government support through commitment to the 863 and 973 Programs, among others, for key S&T areas with direct relevance for future social and economic development, well before China could afford to do so, led to quite a number of breakthroughs. At the turn of the twenty-first century, alongside the United States, Germany, Britain, France, and Japan, China became part of the International Human Genome Project (IHGP). Although its contribution amounted to just 1 percent of the total sequencing of the human genome, China was the only developing country to participate in the consortium. In fact, genomics has also contributed to the development of GM crops in China, with the 863 Program supporting the mapping of a rice genome. The Beijing Genomics Institute, a major Chinese contributor to IHGP, completed the superhybrid rice (*indica*) genome sequence well in advance of the dawdling international rice-genome consortium, which, led by Japan, had many more resources.[54]

It is estimated that China invested more than RMB1.5 billion ($180 million) in the life sciences/biotechnology between 1996 and 2000, and significantly increased that investment to RMB5 billion ($600 million) between 2001 and 2005. The 863 Program alone allocated about 20 percent of the funding to agricultural biotechnology research in the Tenth Five-Year Plan period (2001–2005), which totaled RMB3 billion ($360 million).[55] As a whole, China's gross expenditure on R&D jumped from RMB50.9 billion ($7.5 billion, 0.64 percent of GDP) in 1997 to more than RMB1.57 trillion ($236 billion, 2.11 percent of GDP) in 2016, with expenditure on the life sciences/biotechnology R&D usually accounting for some 20 percent of the total. In 2011, of the RMB51 billion ($7.9 billion) funding by MOST and the National Natural Science Foundation of China (NSFC), RMB9.7 billion ($1.5 billion) was life sciences/biotechnology-related.[56] Right now, in mainland China, as many as 2,500 leading scientists are active in life sciences research at about 500 publicly funded labs in universities and R&D institutes, along with some 7,500 life sciences/biotechnology companies with over 250,000 employees, including many research personnel.[57]

The International Linkages of China's Transgenic Research Community

With a supportive policy environment and the state's commitment in the form of investment, the next most critical element in developing science and technology is talent. In this regard, members of China's life sciences/ biotechnology community have benefited from a particularly strong tradition of overseas training. China's scientific leadership in the Republican era and the early period of the People's Republic overwhelmingly came out of the leading scientific research institutions in the United States, Europe, and Japan. For example, those involved in insulin synthesis were returnees from the United States and Britain; Cornell University saw an early concentration of Chinese students in biology and especially plant genetics, ten of whom would later become elite CAS members.[58]

China's reform and open-door era coincided with globalization of education and research. Consequently, a significant number of Chinese have gone abroad, not only exposed to but also increasingly active at the frontier of international science. The life sciences are among the fields in which Chinese expatriates have formed a critical mass. Life scientists, especially those in the United States, have chosen to stay on to pursue their careers upon finishing their studies, although recent years have witnessed an increasing number returning to China to take advantage of a booming economy and the opportunities available to them.[59] Such international connections have led to high-quality publications by ethnic Chinese life scientists in leading journals such as *Cell, Nature,* and *Science.*[60] Beginning in 1997, international collaboration was clear in publications in agriculture, as well as in neuroscience, behavioral medicine, clinical and experimental medicine, biomedical research, biosciences, and biology.[61] International collaboration in molecular biology intensified in 1999 with the proportion of links with U.S.-based scientists at about 40 percent; internationally collaborative publications have also had a notably higher average impact than indigenously produced ones, measured by citations.[62] It is unclear to what extent such collaboration has involved expatriate Chinese life scientists, but their large numbers—one-third of the approximately 300,000 Chinese students overseas, according to a 2007 estimate,[63] and approximately 2,500 on faculty at American universities, according to another 2007 survey[64] (recent numbers are likely higher)—point to the possibility that they are becoming a strong driver for such collaboration. This expatriate Chinese scientific community is also the talent pool upon which the Chinese government, universities, and other organizations have always wanted to draw.

Returnees' international linkages are evident among transgenic scientists. Some of them embarked on their scientific professions at foreign institutions of learning, most notably Cornell University, Washington University in St. Louis, and the University of California, Davis, in the United States, and the International Rice Research Institute in the Philippines, to name just a few. They are also closely linked to the Rockefeller Foundation and Monsanto.

Among the pioneers and the most important advocates of GM plant research in China is Chen Zhangliang. In 1983, Chen was on a Chinese government scholarship to pursue a Ph.D. at Washington University in St. Louis, where the biotechnology giant Monsanto happens to be headquartered. He worked with Professor Roger Beachy, a pioneer of transgenic technology in the Division of Biology and Biomedical Sciences, on virus-resistant tobacco, the first GM plant, for his doctoral research. Chen received his Ph.D. in 1987 and soon returned to China to start his lab at Peking University, extending his doctoral work on GM tobacco, which in fact became China's first GM crop to commence production in 1992. Chen was appointed vice president of Peking University in 1995, president of China Agricultural University in 2002, vice governor of Guangxi Autonomous

Region in 2008, and vice chair of the China Association for Science and Technology, the umbrella organization of China's professional societies, in 2013. Despite his declining influence after becoming a bureaucrat, Chen Zhangliang continued to be at the vanguard of the pro-GMO camp. Chen was a member of the Standing Committee of the Eleventh and Twelfth CPPCC before resigning in November 2016 for personal reasons.

At Huazhong Agricultural University in Wuhan, Hubei Province of central China, Zhang Qifa, mentioned several times in this chapter, is another strong advocate of transgenic technology. He himself has never researched GM crops; in fact, he did not foresee the prospect of plant molecular biology when he studied plant genetics at the University of California, Davis, where he received his Ph.D. in 1985 and stayed on for a year as a postdoctoral fellow.[65] His research after returning to China has been funded by the Rockefeller Foundation, which at one time also appointed him to the scientific advisory board of its International Program on Rice Biotechnology, as well as other international organizations. Zhang was elected a CAS member in 1999 at age forty-six, the youngest at the time, and a foreign associate of the U.S. National Academy of Sciences (NAS) in 2007. Zhang also has close connections with Monsanto, which in 2009 invested $160,000 in a special scholarship at Huazhong Agricultural University overseen by Zhang for strengthening cooperation between Huazhong and Monsanto in agricultural biotechnology. Zhang has leveraged his international connections in the training of his students by placing them at international centers of agricultural biotechnology. For example, Tu Jumin, one of Zhang's doctoral students, spent three years between 1995 and 2001—first as a doctoral student then as a postdoctoral fellow—at the International Rice Research Institute in the Philippines, where he carried out critical work on the strains of Bt rice that later would be granted biosafety certificates for commercialization.

With support from the Rockefeller and Ford foundations, International Rice Research Institute is often touted as the first, and a leading, international agricultural research center under the Consultative Group on International Agricultural Research (CGIAR), focusing on promoting GM-rice planting. Also hailing from this institute is Huang Jikun, an agricultural economist. Huang received his Ph.D. in economics from the University of the Philippines in 1990, followed by a two-year postdoctoral research stint at the nearby rice institute. As longtime director and chief scientist of the CAS Center for Chinese Agricultural Policy (CCAP), China's leading think tank on agriculture policy, with transgenic technology as one of its focuses, Huang has collaborated with leading international centers for agriculture policy, including the International Food Policy Research Institute at Stanford University, and with leading agricultural economists such as Scott Rozelle at the University of California, Davis (now at Stanford), and Carl Pray at Rutgers, the State University of New Jersey, among many others.[66] He has also advised the World Bank and the UN Food and Agriculture Organization and is on the board of the International Service for the Acquisition of Agri-biotech Applications (ISAAA). Huang moved to Peking University in 2015 but remains as the center's director.

The Cornell University connection continues to be strong. Li Jiayang did his postdoctoral research there after receiving his Ph.D. from Brandeis University in 1991. Best known for his work on Arabidopsis, a model organism, Li returned to the CAS Institute of Genetics in 1994 and became CAS vice president ten years later. He was elected a CAS member in 2001 and a foreign associate of the U.S. NAS in 2011, the same year that he was appointed vice minister of agriculture and CAAS president. At one time, Li served on the advisory board of DuPont Pioneer, also a leading developer and supplier of seeds based on advanced plant genetics. In November 2012, Li was elected an alternate member of the eighteenth CCPCC, entering the inner circle of Chinese politics. Zhu Yuxian, Chen Zhangliang's onetime colleague, received his Ph.D. from Cornell in 1989 and went on to become a postdoctoral fellow at Washington University in St. Louis before returning to Peking University in 1991. Zhu now heads the Academy of Advanced Studies at Wuhan University and is also a CAS member, elected in 2011. Huang Dafang, another scientist active in transgenic research, was a visiting scholar at Cornell between 1986 and 1988 and again in 1992.

Many leading Chinese scientists engaged in GM-crop research also are internationally well connected. Trained in microbiology in the former Soviet Union in the late 1950s and early 1960s, Fan Yunliu once worked at the CAS Institute of Microbiology. In the early 1980s, she spent two years studying molecular biology at the University of Wisconsin and Northwestern University Medical School as a visiting scholar. Upon returning, she moved to CAAS to set up a molecular biology lab, where she developed China's first GM rice in 1989[67] and later phytase maize, which received a biosafety certificate in 2009 along with the two Bt-rice strains. In 1997, Fan was elected a CAE member. Zhu Zhen, a CAS trained Ph.D., did his postdoctoral research at the University of Tennessee. Guo Sandui, the "father of China's Bt cotton," was at the Pasteur Institute in France between 1986 and 1988 to supplement his training as a worker-peasant-soldier student at Peking University during the Cultural Revolution. Jia Shirong, another Bt-cotton inventor, was a visiting scholar at Canada's Plant Biotechnology Institute and University of Saskatchewan from 1979 to 1981. Peng Yufa, a CAAS researcher on plant protection, carried out collaborative research in the United States in 1985 and Australia in 1996.

As in the cases of Zhang Qifa and Huang Jikun, some other leading agricultural biotechnologists have had their research supported by international organizations. Although Wu Kongming—a CAE member, a CAAS vice president, an expert in plant protection, and Peng Yufa's colleague chairing the National Biosafety Committee on Agricultural GMOs—does not have a foreign training background, his research has been supported by the Rockefeller Foundation and the U.S. Department of Agriculture (USDA). Zhu Zhen also benefited from a Rockefeller grant to start his GM research program upon finishing his postdoctoral study in the United States.

Finally, I want to mention Xue Dayuan, an expert on biodiversity with the Ministry of Environmental Protection's Nanjing Institute of Environment Sciences and the Central University of Nationalities in Beijing, and Gu Xiulin, an agricultural economist now affiliated with the Yunnan University of Finance and Economics. Xue, who at one time was deputy director of MOEP's Office of Biosafety, representing the Chinese government in the drafting, negotiation, and implementation of the UN Convention on Biodiversity and related documents, had become a leading advocate of risk assessment of GM crops and biosafety in the last decade or so. After receiving his bachelor's and master's degrees from Nanjing Agricultural University, Xue studied Bt cotton at the University of California, Santa Cruz, in 1988 before pursuing a Ph.D. in geography at Nanjing University. From the end of 1998 to the end of 2000, Xue Dayuan was a postdoctoral fellow in economics at Australia's Queensland University. Gu Xiulin received her master's degree in agricultural economics from the Chinese Academy of Social Sciences (CASS) in 1985. She then spent several years at the CASS Institute of Rural Development before going to the University of Hawaii in the United States, where she earned a Ph.D. in agricultural and resources economics in 1999. Gu is a leading academic and hard-liner in the anti-GMO camp.

Research on and Commercialization of GM Crops

Research

In the last three decades, political support, resource mobilization, and a large internationally well-connected talent pool have pivoted China into a global leading position in the research on and commercialization of agricultural biotechnology. Agricultural biotechnology has focused mainly on plant biotechnology or GM crops with the allocation of significant resources to the 863 Program, the 973 Program, NSFC, and lately, MLP. Internationally, the Rockefeller Foundation, McKnight Foundation, and China–European Union Science and Technology Cooperation Program, among others, have also lent their support to China's efforts.[68] Consequently, the nation has researched a number of GM crops and expanded its agricultural biotechnology program into one of the largest and most advanced in the developing world.

In particular, China's R&D expenditure on plant biotechnology increased from RMB140 million ($41 million) in 1986 to RMB920 million ($110 million) in 1999. During the same period, the number of plant biotechnologists rose from 740 to 1,988.[69] Most significantly, in 1999, the State Council under Premier Zhu Rongji approved the initiation of a dedicated GM Plant Research and Commercialization Special Program (*zhuanjijin zhuanxiang*), allocating public funding of RMB510 million ($62 million) with matching funds of RMB320 million ($39 million) from other sources. By the end of 2006, the program had successfully fulfilled its original tasks and goals by generating major technological, economic, and social impacts. In particular, it led to publication of papers in *Cell*, *Nature*, and the *Proceedings of the National Academy of Sciences*, as well as in journals indexed by *Science Citation Index*, a database compiled by the Institute for Science Information (later Thomson Reuters, now Clarivate Analytics), and *Engineering Index*. The program also attracted the return of Chinese scientists from overseas, saw the gradual coming of age of a cadre of agricultural biotechnologists, and funded the establishment of two national research centers for GM crops in Beijing and Wuhan and two experiment and commercialization bases for plant GMOs—corn and soybeans in Jilin and cotton in Henan.[70] Scientists cloned 418 important function genes and applied for 295 domestic patents. One of the program's significant achievements was an increase of the share of domestically developed Bt cotton from a mere 5 percent in 1998 to 70 percent in 2005 (although this figure is controversial, as will be discussed). Investment in agricultural biotechnology during the Tenth Five-Year Plan period (2001–2005) was increased five times, reaching $500 million by the end of the plan period.

In 2006, the State Council unveiled the fifteen-year MLP as a roadmap for China's becoming an innovation-oriented nation, listing new varieties of GMOs as one of the sixteen Mega-Engineering Programs. The program was officially launched on July 9, 2008, when the State Council's Standing Committee finally gave it the go-ahead after at least thirty consultation meetings. The deliberation was not about whether China should develop transgenic technology but whether and when China should commercialize a particular GM crop and how China should strengthen a stringent biosafety and risk-assessment regime. Following State Council's approval, the discussion shifted to how to accelerate and implement the program.[71] Oriented toward scientific research, the program aims at developing second-generation GM crops with disease-resistance, drought-resistance, higher-quality, and higher-yield traits on top of insect resistance and herbicide tolerance, and promoting efficient and effective agricultural production. The long-delayed biosafety certification process for GM crops was accelerated, ultimately issuing biosafety certificates to Bt rice and phytase maize in 2009. A CCAP survey in 2010 estimated an expenditure of at least $1.2 billion for agricultural biotechnology research in 2009.[72] Between 2009 and 2014, the central government appropriated a total of RMB6.465 billion ($982 million) to MEP (see table 8.1).[73] Agricultural biotechnology was again identified as one of the priorities in the Plan for the Development of China's Strategic Emerging Industries, launched in October 2010, although uncertainties still linger, especially on the commercialization prospects for GM crops.

According to the same 2010 CCAP survey, there has been a concentration of research efforts at some key Chinese universities and public research institutes, where a cadre of 3,400 researchers and administrative personnel in agricultural biotechnology (plants, animals, and microorganisms), of whom some 400 hold Ph.D. degrees, is one of the largest in the world (the number can only have increased further over the last several years). As one of the early pioneers and still an important player, Peking University conducted research on GM tobacco, tomatoes, pepper, and flowers, and set up an enterprise to commercialize its research. CAS is another stronghold of

GMO research, with the Institute of Microbiology also introducing virus-resistant tobacco and carrying out field tests in 1992 while the Institute of Genetics developed several types of insect-resistant Bt rice. At CAAS, more than RMB50 million ($6 million) was devoted to developing bollworm-resistant cotton seeds containing Bt genes, which eventually became China's first commercially viable GM crop. And at Huazhong Agricultural University, research laid the foundation for the biosafety certificates issued to two strains of Bt rice in 2009.

Commercialization

China is also one of the global leaders in the commercialization of GM crops. The 863 Program had a commercialization component, thus giving birth to China's modern biotechnology sector. The first GM crop in China, a virus-resistant tobacco, was developed by Chen Zhangliang, then at Peking University, in the late 1980s, and entered commercial production in 1992; it was exported to the United States to be used in Marlboro cigarettes. At that time, the United States was still debating the safety of GM crops. Concerned that American cigarettes containing transgenic tobacco might be rejected elsewhere, USDA dispatched a delegation from its Beltsville Agricultural Research Center, including Dr. T. C. Tso, the highest-ranking Chinese-American official and scientist, to China for negotiation. The Americans urged China to stop exporting the virus-resistant tobacco to their country, and China eventually removed the transgenic tobacco from production, thus ending its, and the world's, first commercial cultivation of GM crops.[74]

Throughout the 1990s, growth of the agricultural biotechnology program accelerated in China. Between 1997 and 2001, the Ministry of Agriculture approved 504 agricultural GMO biosafety applications, including 316 for field tests, 129 for environmental releases, and 59 for biosafety certificates for commercialization.[75] In the next decade, MOA approved 2,275 applications for field tests, 459 for environmental releases, and 317 preproduction tests, and issued 1,830 biosafety certificates.[76] At one time, China's agricultural research community expected to make China the first country in the world to commercialize GM rice and corn.[77] Thus far, however, MOA has only issued licenses to six GM plants involving ten events[78] for commercial production: virus-resistant tobacco, sweet peppers, and color-altered petunias (Peking University); slow-ripening tomatoes (Peking University, Huazhong Agricultural University, and CAS Institute of Microbiology); bollworm-resistant Bt cotton (CAAS and Monsanto); and virus-resistant papaya (South China Agricultural University). MOA has also issued bio-safety certificates for commercialization of two strains of Bt rice (Huazhong Agricultural University) and herbicide-tolerant phytase maize (CAAS and Origin Agritech) (table 3.3; also see further discussion in chapter 4).[79] In 2016, with 2.8 million hectares devoted to GM varieties, mostly Bt cotton, China ranked eighth in the world, after the United States, Brazil, Argentina, Canada, India, Paraguay, and Pakistan (see table 1.1).

Of the GM crops approved for commercial cultivation, certificates for GM tomatoes have expired and have not been renewed. Apparently, GM tomatoes have low yield, thick skin, and poor taste with no storage advantages, facing challenges similar to those that led to Calgene's withdrawal of GM tomatoes from the market in 1997. GM sweet peppers also do not have obvious advantages over conventional ones. In reality, only Bt cotton and papaya have been grown on a relatively large scale in China; the former, to be discussed in detail in the next section, has resulted from the substantive support of the above-mentioned agricultural biotechnology programs.[80]

China is also the first and only country in the world to achieve the commercialization of GM poplars. After MOA's Agricultural Genetic Engineering Biosafety Committee approved environmental release for Bt European black poplar trees, cultivation began in Manas, Xinjiang autonomous region, in 1998, followed by similar tests in Beijing, Jilin, Shandong, Jiangsu, Henan, and Shaanxi in 1999. The Forestry Biogenetic Engineering Safety Committee under State Forestry Administration approved its commercial cultivation in 2002. In 2001 and 2002, the same committee also approved the environmental release and commercialization of another GM poplar event—bivalent-insect-resistant hybrid poplar 741. As of 2011, a total of 450 hectares of insect-resistant poplars had been planted.[81] However, the certificates have not been renewed since the involved researchers retired.[82]

China has placed its biggest hope for GM crops on rice, the world's most widely consumed grain, with groups of scientists at Huazhong Agricultural University, National Rice Research Institute, the CAAS Institute of Bio-technology, and the CAS Institute of Genetics, among others, working on various varieties. The two strains that Huazhong Agricultural University developed and that were granted biosafety certificates in 2009 were said not

TABLE 3.3

GM Plant Events Approved for Commercial Planting or Issued Biosafety Certificates in China

No.	Crop	Trait	Event	Developer	Year of Approval
1	Tobacco	VR	PK873	Peking University	1992
2	Cotton	IR	GK12	Chinese Academy of Agricultural Sciences	1997
3	Cotton	IR	GK321	Chinese Academy of Agricultural Sciences	1997
4	Cotton	IR	MON531	Monsanto	1997
5	Tomato	DR	Huafan 1	Huazhong Agricultural University	1998
6	Tomato	VR	PK-TM8805R	Peking University	1998
7	Tomato	DR	Dadong 9	Institute of Microbiology, Chinese Academy of Sciences	1998
8	Petunia	FC	Petunia-CHS	Peking University	1999
9	Sweet pepper	VR	PK-SP01	Peking University	1999
10	Poplar trees	VR	European black poplar trees	Institute of Forestry, Chinese Academy of Forestry	2002
11	Poplar trees	VR	Poplar 741	Hebei Agricultural University	2003
12	Papaya	VR	Huanong No. 1	South China Agricultural University	2006
13	Rice	IR	Bt Shanyou 63	Huazhong Agricultural University	2009/2014*
14	Rice	IR	Huahui-1	Huazhong Agricultural University	2009/2014*
15	Maize	HP	BVLA430101	Chinese Academy of Agricultural Sciences and Origin Agritech	2009/2014*

*Events have received biosafety certificates but their commercial planting has not been approved.

Note: DR = delayed ripening/altered shelf life; FC = modified flower color; HP = high phytase; IR = insect resistance; VR = virus resistance.

Sources: Yunhe Li et al., "Biosafety Management and Commercial Use of Genetically Modified Crops in China," *Plant Cell Report* 33 (2014): 570; Carl E. Pray et al., "The Impact of Bt Cotton and the Potential Impact of Biotechnology on Other Crops in China and India," *Frontiers of Economics and Globalization* 10 (2011): 89; Office of Agricultural Transgenic Biosafety Administration Under the Ministry of Agriculture, Institute of Biotechnology Under the Chinese Academy of Agricultural Sciences, and China Society for Agricultural Biotechnology, *Thirty Years of Transgenic Technology* (in Chinese) (Beijing: China Agricultural Science and Technology Press, 2012).

to be the best strains.[83] Research that has attracted equal if not more attention among GM-rice efforts is that carried out at National Rice Research Institute in Hangzhou, Zhejiang Province, as it promises a variety of rice with significantly higher yields, better quality, and resistance to drought and insects than non-GM varieties. In 2001, this GM-rice variety was approved for environmental release and began sales on a test basis in several counties in eastern Zhejiang Province. In addition, scientists have developed rice varieties using bacteria to achieve nitrogen fixation.[84]

Current work includes developing GM crops with resistance to viruses, diseases, insects, bacteria, and fungi, and with herbicide tolerance. GM crops such as wheat, corn (maize), soybeans, potatoes, canola, peanuts, and others have gone through different stages of lab study, field tests, or environmental release, but have not reached the stage of biosafety certification for commercialization.[85] Although biosafety certificates for the two strains of Bt rice and phytase maize expired and were then renewed, Chinese agricultural biotechnologists are still uncertain when they will be able to commercially plant them and when their endeavors will lead to the granting of biosafety certificates for commercialization of other GM crops.

The Bt-Cotton Story

Overview

China is one of the world's largest cotton producers, devoting a significant amount of its land to this important cash crop. Cotton is more vulnerable to insects, especially bollworm, than other crops grown in China. The incidence of cotton bollworms increased dramatically in the early 1990s, devastating cotton farmers. In order to control bollworm attacks, farmers had to apply increasing amounts of pesticide, which then caused health problems. Transgenic cotton was developed and introduced exactly at the time when the future of the crop was in jeopardy.

The Bt-cotton varieties grown in China involve three Bt-gene events developed by the CAAS Institute of Biotechnology and Monsanto. Chinese agricultural biotechnologists started research on Bt's application to cotton in the 1980s. By 1992, Guo Sandui and Jia Shirong at the CAAS Institute of Biotechnology had synthesized the gene Cry1A, a Bt gene, and raised its expression by about one hundred times. In the following year, they incorporated this univalent gene into cotton varieties planted in the Yangtze and Yellow River cotton regions to create China's first-generation Bt cotton, making China the second country in the world, after the United States, to possess intellectual property rights (IPRs) covering both upstream basic research and downstream breeding in GM cotton. They then recombined Cry1A and CpTI, another gene that also provides insect protection, to develop a new variant of bivalent insect-resistant Bt cotton containing both genes. Scientists spent the next several years conducting field tests to demonstrate the effectiveness of the univalent and bivalent genes.[86]

However, skeptical about the indigenous technology, MOA turned to Monsanto. In the mid-1990s, when the Chinese cotton farmland decreased because of the bollworm attack, Monsanto began to carry out required field tests for the introduction of its gene, Cry1Ac, also a Bt gene, with the brand name Bollgard, to cotton in China. But Monsanto asked for a $10 million patent-licensing fee, required its joint venture Jidai in Hebei Province to buy Monsanto Bt-cotton seeds for sale in China, and also requested a 50 percent share of the profits from selling the seeds. Having learned of these terms, Vice Premier Zhu Rongji encouraged Chinese scientists to improve their own technology.[87]

At the end of 1997, MOA approved commercialization of the Bt-cotton events of both CAAS and Monsanto, making cotton the first major GM crop produced on Chinese soil. The rapid and widespread cultivation of Bt cotton in China's cotton-growing regions significantly contributed to China's cotton production.[88]

It took just two years for China to approve the Bt-cotton commercialization, bypassing some required safety tests.[89] Such relative quickness and ease could be attributed to several factors.[90] First, Bt cotton had already been approved in the United States, which China always looks to in developing its technology. Second, Bt cotton did not seem to present a food-safety risk, although cotton oil and meals could enter the food chain; nor did it likely cause agronomic problems.[91] Nevertheless, Monsanto was required to conduct environmental-impact assessments as well as animal tests to ascertain the safety of cotton oil and meals. These tests were purported to buy time for the readiness of the indigenous Bt-cotton seeds.[92] The CAAS scientists also indicated that a mixed cropping tradition used by cotton farmers in North China made it unnecessary to establish refuges for non-Bt cotton. The third and probably most important reason was that the international controversy concerning GM crops had not emerged.

Bt cotton was rapidly adopted by Chinese cotton growers and spread to a significant part of the country, with cultivation area expanding from only 2,000 hectares in 1997 to some 700,000 hectares in 2000, or about 20 percent of China's total cotton acreage.[93] By the early twenty-first century, Bt-cotton varieties had been planted in twelve provinces. In 2002, Bt cotton was grown on 2.2 million hectares, or 53 percent of total cotton acreage, making China the world's top Bt-cotton producer.[94] In 2009, China grew more than 2.8 million hectares of Bt cotton, or about 70 percent of the total cotton production area. In 2015, China further enlarged its Bt-cotton-growing area to 3.7 million hectares, or more than 96 percent of the total cotton production, but this decreased to 2.8 million hectares in 2016.[95]

Competition Between Indigenous and Foreign Genes

At the turn of the century, the CAAS and Monsanto varieties split the Chinese cotton farmland, with each accounting for about one million hectares, but today Monsanto accounts for only about 10 percent of total production. On one hand, the availability of the CAAS Bt genes has enabled their Bt cotton to spread rapidly to compete with the Monsanto gene. While CAAS first worked with Delta & Pine Land (D&PL), an American seed company, to test Bt cotton in 1995, most of its efforts were through a joint venture with Biocentury, a company in Shenzhen that made its initial money in real estate, and later with Origin Agritech, a NASDAQ-listed agricultural biotechnology company set up by returnees from overseas. The two companies marketed and licensed cotton seeds containing CAAS-developed Bt genes to many plant-breeding institutes and seed companies, mostly small- and medium-sized, which then incorporated those genes into their breeding programs. In contrast, Monsanto only authorized its Boll-gard gene to be used by its two joint ventures—Jidai in Hebei and Andai in Anhui Province—formed with D&PL and provincial seed companies.[96]

On the other hand, although the only legitimate Bt-cotton varieties on the Chinese market are those from Biocentury/Origin Agritech or Jidai and Andai, which are approved for use only in particular provinces, many of the seeds grown in China have been generated by a large number of local research institutes and seed companies that backcross commercially available seeds from either CAAS or Monsanto.[97] Some of them, particularly those early on the market, did not even undergo required biosafety-approval processes. The short-term results were largely positive for farmers, who had access to widely available unauthorized low-priced seeds. As the seed market becomes more complex, farmers have had trouble identifying superior varieties and often prefer to save seeds of the varieties that perform well. But as developers have their IPRs violated in an unregulated seed market and have difficulty recouping their investment in R&D, neither CAAS nor Monsanto has sufficient incentives to develop, improve, and introduce new genes.[98]

Furthermore, like other foreign companies, Monsanto was constrained by China's regulatory requirements. Although China proudly claims the domination of its indigenous genes, in reality the Bt-cotton varieties originating from Monsanto's Bollgard gene account for a much larger share than official statistics indicate, simply because many domestic seed companies backcross seeds with the Bollgard gene. Therefore, Monsanto has played a de facto more important role in China's Bt-cotton development and commercialization. Having seen its gene pirated, however, Monsanto stopped introducing new Bt-cotton genes into China, a loss to Chinese cotton farmers as they are unable to access the most advanced technology.[99]

Benefits to Farmers and the Environment

Early assessments of the impact of developing Bt cotton in China appeared to be largely encouraging, including increased yields and reduced pesticide exposure and therefore environmental and health benefits.[100] Recent studies have indicated that Bt cotton continues to control bollworm, and farmers rather than biotechnology or seed companies continue to be major beneficiaries, contrary to the anti-GMO camp's allegations.

First, there is evidence of the positive impacts of Bt cotton in reducing the use of pesticide, lowering farmers' costs, and increasing cotton yields. Farmers growing Bt cotton sprayed two-thirds less pesticide than those who did not adopt the Bt varieties, thus significantly decreasing their expenditure on pesticide. Farmers also cut labor costs, largely because of reductions in the required number of pesticide sprays. Moreover, yields of Bt cotton were between 7 and 15 percent higher.[101] Although the price of Bt-cotton seeds is higher than that of conventional seeds, the difference in seed costs is offset by the reduction in spending on pesticide and labor, thus decreasing overall production costs for Bt cotton and increasing net revenue as compared to non-Bt cotton. For example, between 1999 and 2001, the Monsanto seed was about RMB100 ($12) per kilogram while the CAAS seed was roughly half that, compared with an ordinary seed price of about RMB10 per kilogram;[102] and the amount of Bt-cotton seed required was slightly higher than with conventional seed (3,371 kg versus 3,186 kg per hectare). However, in three northern Chinese provinces, Bt cotton reduced pesticide use by an average of thirteen sprays and by 35.7 kilograms per hectare, a 55 percent reduction of pesticide use. This, in turn, reduced costs by $762 per hectare per season, a decrease of 28 percent, in addition to ensuring sustainable production against bollworms. A study conducted between 1999 and 2007 further confirmed that net revenue from planting Bt cotton exceeded that from growing conventional cotton.[103] Aggregate cotton yields continue to rise in China, suggesting that Bt cotton continues to do well.

Second, the reduced need for highly toxic pesticides substantially reduces the risk and incidence of pesticide poisoning.[104] The obvious benign effects on human health include reducing chronic toxicity, impact on the human nervous system, and hormonal changes, which would otherwise lead to increase incidents of poisoning and even death. Since the introduction of Bt cotton, pesticide toxicity has been reduced to 4.7 percent of Bt-cotton farmers compared to 22 percent of non-Bt-cotton farmers; this reduction has translated into total benefits of $334 million, of which $197 million, or close to 60 percent, could be attributed to the CAAS efforts.[105] The reduction of pesticide use also brings substantial environmental benefits.

Third, the bollworm population in the Bt-cotton-growing area has declined. Data collected from 1997 to 2007 in six northern Chinese provinces indicate that Bt cotton was not only efficacious in controlling bollworm infestation, especially from 2002 to 2006, but also responsible for long-term suppression of the bollworm population, thus benefiting farmers of other crops that are also susceptible to the insect.[106] In Henan Province, where farmers cultivated some non-Bt-cotton fields as late as 2006 and 2007, not only had pesticide use against bollworm in Bt-cotton fields between 1999 and 2007 been less than ten kilograms per hectare for the entire period except 2000, but spraying for bollworm on non-Bt-cotton fields had also declined dramatically since 1999.[107]

Fourth, growing Bt cotton has generated social welfare effects as the introduction of Bt cotton lowers not only the cost of cotton production but also the price of cotton. A cheaper domestic cotton supply leads to an expansion of exports of finished fabrics that sell at a somewhat lower price. Gains for China as a result of using Bt cotton reached $300 million.[108]

Between 1997 and 2006, the head start in Bt cotton brought benefits worth RMB12 billion to China's textile industry and consumers, on top of a RMB25 billion gain for its Bt-cotton farmers. The return in 2006 alone reached an overall economic benefit of RMB6 billion ($750 million), more than enough to offset all the expenditure that China had invested on the research on and commercialization of all GM crops.[109]

Summary and Discussion

In most of the past three decades, China's development of biotechnology in general and transgenic technology in particular has benefited from a promotional R&D policy environment. With the reform of its S&T system encouraging the application of research on the economy and academic entrepreneurship, the quality improvement of the talent pool, and the return of a contingent of scientists trained at leading international centers for agricultural biotechnology and experience in industry and startups, China started to develop transgenic technology and GM crops at almost the same time as other GM nations. Chinese scientists have simultaneously played the role of policy entrepreneurs, involved in policymaking pertaining to their research, and of practitioners in the commercialization of their research. The Chinese leadership, especially former Chinese premiers Zhu Rongji and Wen Jiabao and current leader Xi Jinping, has also recognized the enormous potential of biotechnology. All these factors have explained the channeling of public funds through a series of dedicated programs, especially the 863 Program and MLP's Mega-Engineering Program for the cultivation of new GMO varieties. Meanwhile, attention from the political leadership and enormous investment of public funds in developing GM crops have put pressure not only on scientists but also on government agencies that decided to make the investment initially. This orientation toward quick and immediate returns has inevitably motivated scientists and bureaucrats to move forward on commercialization, which has also been one of the critical goals of the reform of the S&T system. Although China has only approved Bt cotton as a major crop for commercial cultivation, despite what many scientists had originally envisioned, the efforts on the development of Bt cotton seem to have paid off, tilting the debate on Bt cotton toward benefits. Nonetheless, the agricultural biotechnology research community has been dissatisfied with having only one extensively cultivated GM crop.

The development of transgenic technology in China has been the outcome of close linkage between domestic agricultural biotechnologists and their colleagues at the frontier of international research. Indeed, international connections became necessary conditions for some of those involved in the research to advance their professional careers, and even to have political opportunities in China, at a time when the nation was dependent upon foreign institutions of learning for the supply of high-end talent.[110] Some scientists have accumulated knowledge and social and political capital by making good use of such connections. But such connections can also prove liabilities, as the same scientists have also been subjected to groundless, politically motivated attacks and humiliations precisely because of such connections, as I will discuss in chapter 5.

CHAPTER IV

Science, Biosafety, and Regulations

A
s discussed in chapter 3, for most of the time since the mid-1980s, China's policy toward agricultural GMOs has been promotional, or at least permissive, reflected in government's strong commitment to funding research on developing GM crops and to promoting their commercialization. Supported by the government and in alignment with the policy, the research community has been mobilized to take initiatives to fulfill the government's objectives, as well as their own, to turn China into a GM nation.

Meanwhile, the Chinese government has been closely monitoring the global debates on GM crops and foods regarding the risks and associated issues, as reviewed in chapters 1 and 2. Pushback from domestic stakeholders has prompted the government to respond to skeptics and critics by gradually establishing and strengthening a biosafety regulatory regime, which by name and nature tends to be precautionary, to balance its promotional stance on research and commercialization.[1] There also is division of labor and even bureaucratic competition between the scientific bureaucracy, whose mission in promoting scientific research and commercialization seems to be clearly defined, and other government agencies, particular of agriculture, environmental protection, and food safety, which have to deal with challenges of a more complex nature.

This chapter begins with an examination of China's evolving regulations germane to GM crops and foods and the rationales behind the setting up and enhancing of such a regulatory regime. It will also discuss the implications of a stringent regulatory regime for risk assessment and management in the context of globalization and China's evolution into a regulatory state. Finally, the biosafety certification of GM rice will be analyzed in detail.

Establishing and Enhancing a Biosafety Regulatory Regime in China

Evolution of the Regime

The 1990s witnessed extremely rapid development in scientific research as well as field tests of GM crops in China, primarily due to a "distinguished feature," or "the level of state investment and volume of investment in the governmental sector."[2] Indeed, China's agricultural biotechnology was almost entirely sponsored by the government, with research carried out mainly at public labs. Early on, the development was mostly unregulated, or there was little scientific or bureaucratic awareness of the significance of regulation, which explains why at one time virus-resistant tobacco could have been grown on a large scale without proper authorization.

It was only in December 1993, seven years into research on GM crops, that the State Science and Technology Commission (SSTC), predecessor of the Ministry of Science and Technology (MOST), issued a "Regulation on the Administration of the Safety of Genetic Engineering," thus giving birth to China's regulatory framework on biosafety (table 4.1). Mostly covering issues of biosafety relating to scientific research, this policy document outlined the general principles of biosafety as well as its categories, risk assessment, biosafety application and approval processes, control principles, and legal responsibilities.[3] Meanwhile, SSTC asked other government agencies to roll out measures on work within their respective jurisdictions. In November 1996, the Ministry of Agriculture (MOA) promulgated "Measures on the Implementation of the Safety of Agro-Bioengineering" to enforce SSTC regulation through delineating the procedures for research, field tests, and commercialization of GM crops. In general, approval was mandatory for doing research on agricultural GMOs, testing them outside of labs and greenhouses, and especially commercialization. Despite including neither labeling requirements nor restrictions on imports or exports of GM products, this MOA document did stipulate the formation of a

TABLE 4.1
China's Evolving Biosafety Regulatory Regime

Date Document Issued	Government Agency Issuing the Document	Title of Document	Development
December 24, 1993	SSTC	Regulation on the Administration of the Safety of Genetic Engineering	Implemented December 24, 1993
July 10, 1996	MOA	Measures on the Implementation of the Safety of Agro-Bioengineering	Implemented July 10, 1996
March 20, 1997	State Council	Regulation on the Protection of New Varieties of Plants	Implemented October 1, 1997
July 8, 2000	NPC's Standing Committee	Seed Law	Implemented December 1, 2000; revised August 8, 2004
May 23, 2001	State Council	Regulations on the Administration of Biosafety of Agricultural Genetically Modified Organisms	Implemented May 23, 2001; revised January 8, 2011 and October 2017
January 5, 2002	MOA	Administrative Measures on the Safety Evaluation of Agricultural Genetically Modified Organisms	Implemented March 20, 2002; revised in 2004, 2016 and 2017

Date	Agency	Regulation	Notes
April 8, 2002	MOH	Administrative Measures on the Safety of Imported Agricultural Genetically Modified Organisms; Administrative Measures on the Labeling of Agricultural Genetically Modified Organisms	First implemented March 20, 2002; revised in 2004 and 2016
		Administrative Measures on the Safety of Foods Based on Genetically Modified Organisms	Implemented July 1, 2002; replaced by Administrative Measures on the Safety of Foods Based on New Resources on December 1, 2007
2002, 2004, 2007, 2011, 2014, and 2016	SDPC, SETC, NDRC, and MOFCOM	Catalogue Guiding Foreign Investment in Industries	From 2016 on, GMOs are no longer a prohibited area for foreign investment
July 2, 2007	MOH	Administrative Measures on the Safety of Foods Based on New Resources	Implemented December 1, 2007; revised as Administrative Measures on the Safety Evaluation of New Food Materials
February 28, 2009	NPC's Standing Committee	Food Safety Law	Implemented June 1, 2009 and revised on April 24, 2015
July 12, 2013	NHFPC	Administrative Measures on the Safety Evaluation of New Food Materials	Implemented October 1, 2013

Notes: The names of the government agencies are the ones used when the programs were initiated: MOA = Ministry of Agriculture; MOH = Ministry of Health; NDRC = National Development and Reform Commission; NHFPC = National Health and Family Planning Commission; NPC = National People's Congress; SDPC = State Development and Planning Commission; SETC = State Economic and Trade Commission; SSTC = State Science and Technology Commission.

National Agricultural Bioengineering Safety Committee to provide the ministry with expert advice on biosafety.[4] In 1997, MOA formed such a committee, which was renamed the National Biosafety Committee on Agricultural Genetically Modified Organisms (NBCAGMO) in 2001, in accordance with a new State Council policy, to be in charge of the risk assessment and biosafety evaluation of GM crops and the administration of mandatory approval procedures for the testing of such crops.[5]

Symbolically, SSTC and MOA documents marked the beginning of the establishment of China's biosafety regulatory regime. In fact, it took more than 120 experts six years to draft the MOA implementation measures,[6] possibly as long as it took to formulate the SSTC regulation. Within China's administrative hierarchy, as noted in chapter 3, a ministerial-level policy has no jurisdiction over activities outside the issuing ministry; such a policy is formulated from the perspective of an individual ministry to represent the ministry's interests. Given its mission in the administration of scientific research, SSTC put forward a regulation on biosafety management in genetic engineering research that was largely promotional; MOA, on the other hand, adopted permissive measures "focused on demonstrated rather than unknown risk . . . and did not assume that uncertainty was itself a risk,"[7] despite maintaining that biosafety was not its responsibility. Such a difference in their respective bureaucratic positions toward GMOs has been persistent, with MOA in charge of the risk assessment and biosafety management process and the science bureaucracy proactively exerting influence.[8] Taken together, a nationwide policy seemed both necessary and urgent.

On May 9, 2001, the State Council stepped in to decree a new policy, "Regulations on the Administration of the Biosafety of Agricultural Genetically Modified Organisms," to replace the early SSTC regulation and to guide official approval of GMOs from research through production, marketing, and trade. The drafting of the regulations involved SSTC, MOA, and the State Environmental Protection Administration (SEPA), predecessor of the Ministry of Environmental Protection (MOEP) and now the Ministry of Ecology and Environment.[9] In particular, the regulations divided biosafety assessment of GM crops into five stages: laboratory research, restricted field test (small scale under contained/controlled conditions), environmental release (medium-scale evaluation in the field), productive field test (a newly added large-scale test in the field carried out before commercial production), and application for a biosafety certificate. Consistent with internationally accepted risk-assessment procedures, the regulations also stipulated a comparative risk assessment approach, under which a GM crop is compared to the corresponding non-GM crop for assessment of environmental/ecological safety and food safety.[10] In addition, they authorized the establishment of an interministerial mechanism on GMO biosafety administration and required multiministerial approval for the commercialization of GM crops. The regulations were most significant in introducing the requirement of mandatory labeling on GM products for retail sale.[11]

Meanwhile, the State Council further instructed MOA to revise its measures to align them with and implement the State Council regulations and especially to tighten its policy in the areas of biosafety management, trade, and most importantly, labeling of GM products. On January 5, 2002, MOA issued a set of three documents—"Administrative Measures on the Biosafety Evaluation of Agricultural Genetically Modified Organisms," "Administrative Measures on the Labeling of Agricultural Genetically Modified Organisms," and "Administrative Measures on the Safety of Imported Agricultural Genetically Modified Organisms"—to replace its earlier measures. These three new measures, which took effect in March 2002 and have been revised several times, greatly expanded the scope of regulations to include more detailed rules on the administration of biosafety, trade, and GM-food labeling, new processing regulations for GM products, and local and provincial-level GMO monitoring guidelines.[12] Most significantly, labeling must be applied to both imported and domestically produced agricultural GM products. Therefore, not only would approved imported GM soybeans and edible oil produced from processed GM soybeans be labeled accordingly, even a food product that used GM-soybean oil in its manufacturing process but contained no trace of GMOs would still have to be so labeled.[13] The measures also stipulated that a GM crop would be approved for import only if it was approved for commercial use in its country of origin.[14]

In retrospect, despite containing precautionary measures, the initial SSTC and MOA policies were promotional or at least permissive. That is, SSTC and MOA introduced policies to show government's genuine interests in advancing agricultural biotechnology and especially commercializing GM crops. The first MOA measures in particular were perceived to be similar to the U.S. GMO biosafety regulations outlined by the Office of Science and Technology.[15] However, with the introduction of the State Council regulations and the modified MOA measures, or with a more comprehensive biosafety regulatory framework, especially the measures on labeling, in place, China imposed the world's most precautionary measures on the production and importation, if not on research, of GM crops, approaching the preventive position taken by the European Union.[16] Thus, the early product-based approach had given way to a biosafety regulatory regime focusing on the entire process from research, environmental release, preproduction test, imports, and exports to the sale of GM varieties, examined case by case.

The Rationales

These regulations and measures might not have been formulated from the spontaneity and consciousness of China's scientific and policy communities as their American counterparts did in the 1970s in developing recombinant DNA technology. Nor might they have shown bureaucratic awareness of the necessity and importance of biosafety and its regulations. Instead, the moves had been passive and strongly influenced by international developments, which is understandable as China was an inexperienced newcomer to biosafety.[17] The first such development was the global regulatory rift regarding GM crops. Within the EU, NGOs such as Greenpeace, some members of the research community, and even some member states started to amplify the potential risks associated with GMOs, thus leading to and maintaining a moratorium on the commercialization of GMOs within its borders. The EU also turned preventive on GM imports.

Second, around that time, a global biosafety protocol was being formed. In 1988, the United Nations Environment Program (UNEP) explored the need for a Convention on Biological Diversity (CBD), which became effective on December 29, 1993, with biosafety as one of the important components. Afterwards, the international negotiation on CBD's implementation, conducted under UNEP, focused, among other issues, on potential threats of GMOs to biodiversity, the environment in general, and human health. By the Fifth Conference of the Parties to CBD, held in Cartagena, Colombia, in 2000, an international protocol on biosafety had been developed. China joined CBD in 1992 and signed and ratified the Cartagena Protocol on Biosafety in 2000. Then, in late 2001, China was admitted into the World Trade Organization (WTO). Despite not requiring strict country regulations on GMOs, WTO membership does stipulate uniformity and nondiscrimination in the application of national standards.[18] In the process of becoming a responsible stakeholder in international affairs, China had acquired experience from foreign countries in legislation, including that regarding the biosafety of GMOs.

Third, there was conflicting information on whether unwanted traits in GM crops would be introduced into conventional crops grown in neighboring farmlands and whether new strains of super-weeds would possibly be created. At the same time, the international scientific community had not reached consensus on the long-term effects of GMOs on human health.

Meanwhile, domestic considerations also came into play. At first glance, one may gather that biosafety regulations were formulated to protect and promote domestic research and commercialization efforts that had already been advanced. Indeed, the State Development and Planning Commission, predecessor of the National Development and Reform Commission (NDRC), and the Ministry of Commerce (MOFCOM) had issued the "Catalogue Guiding Foreign Investment in Industries," banning foreign investment in R&D on GM crops to buy the nation time to nurture its own research endeavor and agricultural biotechnology industry.[19] Also, in the early twenty-first century, Bt cotton was still the only GM crop allowed to grow, and the competition between seeds introduced from Monsanto and developed by the Chinese Academy of Agricultural Sciences was deadlocked. Although the research community was concerned that if China failed to make progress the country would lag behind in biotechnology, some of its members approached transgenic technology cautiously and adversely. China's leading agricultural biotechnologists were in fact on the defensive.

Second, while globally scientists, environmentalists, consumers, and journalists, among others, seemed to have difficulty reconciling their positions on GM crops and foods, the safety of GMOs became a particular concern in China as GMOs were already present in the diets of many Chinese. Chinese started to consume edible oils manufactured from imported GM soybeans in the late 1990s. Under these circumstances, labeling foods with GM contents and giving consumers the right to know and to choose, which will be discussed further in chapter 5, seemed an appropriate step to take. As well as curbing the rush to introduce GM crops and foods, regulations and measures, therefore, would also allow China to wait and observe.

Third, China established its biosafety regulatory regime in anticipation of the WTO accession, after which it would be under pressure to defend against imports of GM crops.[20] Given that the policy was changed amid trade disputes with the United States, Argentina, and Brazil over imports of GM soybeans, China's heightening its precautionary position would also mean undermining those imports and protecting domestic soybean farmers, who numbered about 50 million.[21] Because of the characteristics of GM soybeans—higher productivity per hectare (U.S. average of 2,561 kilograms, a world average of 2,206 kilograms, against China's 1,675 kilograms), higher oil-production yield (18 percent for GM soybeans versus 13 percent for China's non-GM soybeans), and lower price (even after considering transportation cost)—and also because of a surge in demand from domestic producers of edible oils and animal feed, China had witnessed a flood of soybean imports over ten years. In 1993, China produced 15.32 million metric tons of soybeans and imported 99,000 tons; ten year later, domestic production capacity remained largely unchanged while imported soybeans had risen to 20.74 million metric tons, mostly genetically modified soybeans from the United States, Argentina, and Brazil. No import quotas were placed on soybeans, as they were on corn, and wheat, because soybeans are not a staple crop. The dilemma was that China wanted to limit soybean imports but depended on them. When the old trade rules were phased out, biosafety regulations, especially labeling, seemed a rational choice, although under the WTO rules these also are considered a disputable trade barrier. Therefore, when visiting U.S. government officials, from Secretary of State Colin Powell to President George W. Bush, complained about biosafety regulations and labeling of GM products, Chinese premier Zhu Rongji rebutted by saying that China had learned the practice from the EU, Japan, and Korea.[22]

Lastly, the trade issue was further complicated because of the fear that China could lose its advantages in agricultural exports. Although China's exports of products with live modified organisms (LMOs) to Europe were and still are negligible, Europe is a large market for Chinese processed foods. Therefore, European resistance to GMOs was also factored into the tightening of China's biosafety regulations. By self-imposing a "green" trade barrier for GM crops, China might be able to better position its food and agricultural exports as GM-free and retain the option of exporting foods to Europe and Japan, many of whose consumers reject GM foods.[23]

In sum, regulating agricultural GMOs was not simply a matter of regulation per se. China's strategy appeared to have several objectives in mind: encouraging indigenous efforts in R&D and commercialization of GM crops to compete domestically and internationally, restricting imports of GM crops while promoting GM-free crops for exports. However, in gradually tightening its regulations on GM crops, China has slowed its pace as an early leader in the commercialization of GM crops.[24]

Operation of the Biosafety Regulatory Regime

China's biosafety regulatory regime operates through four components at the national level. The first is the Joint Ministerial Conference (JMC), a mechanism within the State Council responsible for overall GMO biosafety administration and coordination as stipulated in the 2001 regulations. Convened by the minister of agriculture and rural affairs (MOARA) once a year, JMC now consists of ministers or vice ministers from eleven other government agencies within the State Council—NDRC; the National Health Commission (NHC), which replaced the National Health and Family Planning Commission (NHFPC) in the 2018 government reorganization; the ministries of science and technology (MOST), education (MOE), finance (MOF), commerce (MOFCOM), and environmental protection (MOEP), now ecology and environment (MOEE); the state administrations for industry and commerce (SAIC), now state administration for market regulation (SAMR), forestry (SAF), now forestry and grassland, and food and drug (CFDA); and the General Administration of Quality Supervision, Inspection and Quarantine (GAQSIQ), with officials of relevant bureaus at these agencies serving as liaisons.[25] JMC coordinates major issues and makes decisions on the biosafety of agricultural GMOs. It also examines and approves major policies and regulations related to commercialization, labeling, imports, and exports of agricultural GMOs.[26]

The second component is MOARA, the successor of the Ministry of Agriculture (MOA), which leads the biosafety regulation by functionally carrying out nationwide oversight and inspection of biosafety involving agricultural GMOs. In 1996, MOA set up an Office of Agricultural Genetic Engineering Biosafety Administration (OAGEBA) to handle day-to-day operations and routine matters. The most important vehicle through which the ministry supervises research and commercialization activities in agricultural GMOs is NBCAGMO.

The third component is the health commission, the successor of NHFPC and Ministry of Health (MOH), in charge of safety management of GM foods. MOH issued the "Administrative Measures of Safety of Genetically Modified Organisms-Based Foods" in 2001 and replaced it with the "Administrative Measures of Safety of New Resources-Based Foods" in 2007. Now, NHC sponsors an appraisal committee, assembling experts on food safety, nutrition, and toxicology at the Chinese Centers for Disease

Control and Prevention (CDC) to review and assess GM crops and foods. The commission then feeds outcomes of the assessment to MOARA to be considered for safety approval of GMOs. Meanwhile, CFDA is charged with safety management of food production and marketing, including labeling of GM foods, while GAQSIQ is responsible for safety issues related to imports and exports of agricultural GMOs.[27]

Finally, the role of MOEE, a successor of the State Environmental Protection Administration (SEPA) and MOEP, is not less important.[28] Given its representation of China in the international negotiation on the UN Convention on Biodiversity and the Cartagena Protocol on Biosafety, SEPA was actively and aggressively involved in the risk assessment of GMOs at least until 2005.[29] It was more than sympathetic to Greenpeace's view on GMOs;[30] in fact, through its Nanjing Institute of Environmental Sciences and the institute's chief scientist Xue Dayuan, an expert on biodiversity who at one time was deputy chair of MOEP's National Biosafety Office, the administration cooperated with Greenpeace soon after it opened an office in Hong Kong in 1997. In June 2002, Greenpeace and the Nanjing institute released a joint report that claimed that Bt cotton would have negative effects on biodiversity, which sent Monsanto's stock tumbling.[31] Between 2004 and 2013, Xue Dayuan and the Nanjing institute partnered with the Third World Network, an international NGO involved in issues relating to development, developing countries, and North-South affairs, to convene five international workshops on risk assessment and management of GMOs in China. Xue became a de facto spokesperson of SEPA and later MOEP, voicing the agency's positions on and concerns over biosafety associated with GMOs. MOEP was also behind State Council's investigation of illegal cultivation of GM crops between 2009 and 2010.

At the provincial level, the responsibility for supervision and administration of the biosafety of agricultural GMOs falls on bureaus of agriculture, which had established provincial biosafety administrative offices by 2005. These offices gather statistics on GM crops tested in the provinces and monitor the performance of research and commercialization. They also screen all applications for GM research, field tests, and commercialization, and disapprove those that they find not up to regulations. Only applications approved by a provincial agricultural bureau may be submitted to the national biosafety committee, thus giving provinces a strong role to play in China's biosafety regulatory regime.[32]

The National Biosafety Committee

One of the most important players in China's biosafety regulation is the National Biosafety Committee on Agricultural Genetically Modified Organisms (NBCAGMO). Convened by MOARA with its members appointed by the ministry on the recommendation of the constituting government agencies of JMC, the committee is responsible for risk assessment and biosafety evaluation, technical advice, and technical guidance for research, field tests, environmental release, and commercialization of GM crops. The term of committee members is three years subject to renewal, which seems to be unlimited. The committee met twice a year when first established as the Agricultural Genetic Engineering Biosafety Committee in 1997. Since 2002, the renamed NBCAGMO meets three times a year to assess applications for biosafety assessment at various stages. An approved application has to receive three-quarters of the votes of the committee's members. Based on its assessment, the committee recommends to OAGEBA in MOARA for approval of activities. Although OAGEBA grants a final approval, NBCAGMO plays the primary role in biosafety management. Biosafety certificates and commercialization approvals are issued on a provincial basis; or the certification is restricted to the provinces that are identified in the application. The certification is valid for five years and usually renewable. In the cases of Bt rice and phytase maize, although NBCAGMO authorized biosafety certification, the final decision might have come from MOA, JMC, or even the State Council.

There have been five national biosafety committees thus far (table 4.2). The first committee, with fifty-seven members, was formed on July 8, 2002.[33] When the second committee was founded, on June 22, 2005, its membership was enlarged to seventy-four. In December 2009, the third NBCAGMO, consisting of sixty experts, was established. In May 2013, sixty-four members were appointed to the fourth committee. The fifth and most recent committee, formed in August 2016, has seventy-five members. Overall, ten members have sat on all five committees, another ten have been members of four committees, thirteen have been on three committees, forty have been on two committees, with the rest serving one term. Such a composition is understandable for maintaining the continuity of the committee's work.

TABLE 4.2

Composition of China's National Biosafety Committee of Agricultural GMOs

	First		Second		Third		Fourth		Fifth		Total	
	N	%	N	%	N	%	N	%	N	%	N	%
Plant molecular biology	8	14.04	12	16.22	14	23.33	12	18.75	16	21.33	62	18.79
Crop genetic breeding	6	10.53	8	10.81	3	5.00	3	4.69	4	5.33	24	7.27
Animal genetic breeding	8	14.04	8	10.81	5	8.33	4	6.25	6	8.00	31	9.39
Veterinary	3	5.26	3	4.05	3	5.00	5	7.81	4	5.33	18	5.45
Microorganism	1	1.75	2	2.70	1	1.67	1	1.56	4	5.33	9	2.73
Ecology	5	8.77	7	9.46	5	8.33	6	9.38	7	9.33	30	9.09
Plant protection	3	5.26	3	4.05	5	8.33	8	12.50	7	9.33	26	7.88
Inspection and quarantine	4	7.02	9	12.16	11	18.33	7	10.94	10	13.33	41	12.42
Food safety	5	8.77	7	9.46	8	13.33	10	15.63	10	13.33	40	12.12
Medicine	0	0.00	0	0.00	2	3.33	2	3.13	7	9.33	11	3.33
Economy and trade	2	3.51	1	1.35	0	0.00	0	0.00	0	0.00	3	0.91
Official	12	21.05	14	18.92	3	5.00	6	9.38	0	0.00	35	10.61
Total	57	100	74	100	60	100	64	100	75	100	330	100

Composed mainly of experts in research and biosafety, as well as administrators, NBCAGMO provides science-based opinions for decision-making.[34] While the number of committee members has changed, the committee has remained stable in terms of the representation of various interests. Most significantly, members from ecology, plant protection, inspection and quarantine, food safety, and medicine have been on the rise and the percentage of officials has been declining. In fact, there have been no members representing the interests of trade and economy and no government officials on the last two committees.

The number of organizations forming NBCAGMO has increased, from twenty-three in the first committee to thirty-four in the fifth one, suggesting a widening representation in this national biosafety regulator. The more recent committees also tend to have more members from provincial organizations, such as academies of agriculture (Heilongjiang, Hubei, Jilin, Shandong, and Sichuan in the third committee; Jilin, Shandong, and Sichuan in the fourth committee; Fujian, Henan, Shandong, and Sichuan in the fifth committee), centers of disease control and prevention (Tianjin in the third and fourth committees; Guangdong in the fifth committee), food- and drug-safety related organizations (Anhui, Guangdong, Hebei, and Shanghai in the third committee; Beijing in the fifth committee), and bureaus of inspection and quarantine for imports and exports (Shenzhen in the third committee; Shanghai in the fourth committee; Shanghai and Shenzhen in the fifth committee).

However, NBCAGMO, which represents the central government overseeing the biosafety of agricultural GMOs, has faced criticisms from its very beginning. First, there was a lack of transparency regarding its membership. The names of those on the first two committees were not disclosed during their tenure. The list of those appointed to the third committee became public only after Wei Rujia, an attorney in Beijing, made a freedom-of-information request to MOA in early 2010.[35] MOA disclosed members of the fourth biosafety committee to the public as soon as it formed the committee; and it put a list of candidates for membership of the fifth committee online to solicit public comments, thus constituting a significant improvement in transparency of the committee's operation.

Second, the national biosafety committee has been criticized for an "unbalanced" disciplinary composition of its membership. For example, Greenpeace alleged that two-thirds of the members were agricultural biotechnologists and their allies, and quite a number of them were from agriculture-related organizations, so that the biosafety evaluation and approval process was deemed biased.[36] At first glance, the committee seems to have been overwhelmed by scientists involved in research on GM crops or having interests in their development, which makes it difficult to rebut that the process might be one in which scientists apply for, assess, and approve applications for various stages from research to biosafety certification to commercialization. The allegation was that as scientist members of the committee perform the role of both "athletes" and "referees," to use a sports analogy, conflicts of interest corrupt the process. Jia Shirong, a three-time committee member, was a target for his multiple identities. Jia is not only a transgenic scientist but also the chief scientist and a board member of Biocentury, a Bt-cotton start-up that he helped establish that uses his technology, so he has commercial interests. Therefore, the committee has been perceived as a club where all scientist members have business interests in the technology assessed. However, approval of Jia's Bt cotton preceded the establishment of NBCAGMO. When his Bt rice was up for consideration in the committee, Jia recused himself to avoid the appearance of a conflict, as did other committee members when applications from their organizations were discussed and deliberated.[37] Most recently, when MOA solicited comments on candidates for the fifth NBCAGMO, Cui Yongyuan, the CCTV anchor turned GMO activist, filed a conflict-of-interest complaint against Lin Min, a four-time member and director of the CAAS Institute of Biotechnology. Lin was removed from membership.[38]

A close look at NBCAGMO's disciplinary composition is probably more revealing (see table 4.2). Indeed, the selection of members for an expert committee that advises policy should consider "qualifications, experience, and judgment" of the members, while giving paramount importance to scientific credentials, as well as "balance and breadth in the appropriate professional disciplines."[39] In the case under study, there has been an almost even split between experts on research (plants, animals, and microorganisms) and on biosafety (ecology, plant protection, inspection and quarantine, food safety, and medicine). Those in the disciplines of plant science and technology can be further categorized into molecular biology and breeding, whose disciplinary focuses are related but not identical. The upstream is basic research, although plant molecular biologists such as Jia Shirong also have been involved in breeding at the downstream.[40] In the third committee, which recommended issuing biosafety certificates to Bt rice and phytase maize, 51.7 percent of the members were on the biosafety side, with plant scientists accounting for 28.3 percent. This committee also had fewer government officials: of the three, two were MOA affiliated—one from the Science and Education Bureau, one from its Center for S&T Development where OAGEBA is located—and the third was from the China Society of Agricultural Biotechnology, a professional society. While Xue Dayuan, an expert on biodiversity and biosafety affiliated with MOEP, had several times been denied inclusion in the national biosafety committee, which some attributed to his connection with Greenpeace, the main reason was his Canadian permanent resident status.[41]

Quite a few members whose disciplines fall into the broad biosafety category are affiliated with CAAS or China Agricultural University, but this does not necessarily imply that they are under institutional obligation or coercion to stand with their colleagues in the scientific disciplines. Therefore, it is unfair to accuse Peng Yufa, a five-time member who has supported Jia Shirong at committee meetings, of conflict of interest just because both are from CAAS and have coauthored papers. Challenging decisions of the committee on such grounds is a typical case of guilt by association.

The third criticism of the national biosafety committee is that neither the State Council regulations nor the agricultural ministry measures stipulate public participation in the biosafety assessment process. Public participation in policymaking has been advocated but not yet practiced widely in China. Indeed, the terms of reference of NBCAGMO fail to specify the importance of public participation in the risk assessment of GM crops. The State Council regulations also neither mandate that public comments be solicited before important approvals nor provide for a period of public examination after the approvals. But in his online interview with *People's Daily*, Peng Yufa indicated that consultative meetings, seminars, and focus groups of various sizes had been held to solicit views from experts, media, the public, and companies on Bt rice and phytase maize, and on the scientific monitoring of Bt cotton. Peng also stated that some outcomes of the consultations had been made available to the public through various channels.[42] However, the issuance of biosafety certificates for commercialization of the two strains of Bt rice in 2009 did lead to a crisis regarding the way in which the agricultural ministry announced such an important policy decision.

Biosafety Certification for Bt Rice

The Announcement

As discussed in chapter 3, the process of regulatory approval of Bt cotton in China was relatively quick, taking just two years for field tests and environmental-impact assessment, although more tests were conducted after its commercialization in 1997. However, the biosafety certification process for major food crops has been extremely slow. According to a 2004 survey by the Center for Chinese Agricultural Policy under the Chinese Academy of Sciences (CAS), the average time for a GM-rice variety to go through the process from applied research to preproduction test was more than eight years.[43] In August 2009, the National Biosafety Committee on Agricultural GMOs finally recommended issuing biosafety certificates to two Bt-rice varieties—Huahui 1 and Bt Shanyou 63—developed at Huazhong Agricultural University (a biosafety certificate also was recommended issuing to phytase maize). The certificates authorized the developer to plant the strains in Hubei Province for five years starting from August 17, 2009. The move was significant not only because these biosafety certificates were the first that China had issued for GM crops since 1997, but also because China consumes and produces the largest amount of rice in the world and rice accounts for 40 percent of the country's total grain output. Therefore, this approval would likely lead to commercial production, which, if followed by other rice-growing nations, could ultimately ease the way for the worldwide adoption of GMOs. At one time, it was expected that commercialization approval would arrive within three to five years after further tests, a requisite process of verifying value for cultivation and use (VCU), and obtaining a new seed variety registration and a production license, required by China's Seed Law. In a word, the planting of Bt rice would start soon in China.

However, such a significant decision was not announced in a serious and transparent manner. MOA did not make a public statement but buried the certification deep in the website of OAGEBA, China Biosafety, on October 22, 2009. Even Greenpeace China, which regularly monitors the website, did not immediately catch this development.[44] The announcement was first picked up by Clive James, chair of the International Service for the Acquisition of Agri-biotech Applications (ISAAA), an international NGO that promotes crop biotechnology. In the December 4, 2009, issue of *Crop Biotech Update*, an ISAAA online newsletter, James highlighted that China's approval "will likely result in a positive influence on acceptance and speed of adoption of biotech food and feed crops in Asia, and more generally, globally, particularly in developing countries."[45]

GM Rice: Research and Biosafety Certification

The granting of biosafety certificates to Bt rice in 2009 was not entirely unexpected. With the successful introduction of Bt cotton, Chinese agricultural biotechnologists had foreseen for years that GM rice would be next in the pipeline for commercialization.[46] There had also been repeated speculations, pronouncements, or media reports that China was very close to approving GM rice for commercialization.[47]

It is estimated that in the past two decades China has devoted some one-third of the public funding on GM crops to the development of GM rice. In 2004 alone, the amount of public investment reached some RMB500 million ($60 million).[48] Research has been carried out at the leading agricultural biotechnology labs at CAAS, CAS, Huazhong Agricultural University, Zhejiang University, and others, focusing on different lines of insect resistance and traits such as herbicide tolerance, increased yield, improved quality, salt and drought tolerance, and biopharmaceutical usages.[49] The first study of insect-resistant Bt-rice transformation was reported in 1989 by CAAS's Fan Yunliu, whose phytase maize was granted a biosafety certificate alongside Huazhong Agricultural University's Bt rice. From January 1995 to December 2009, a total of 378 and 108 peer-reviewed papers, including master's and Ph.D. theses, on Bt rice were published in Chinese and English, respectively.[50] In fact, as of 2005, more than a hundred GM-rice varieties had been in field tests. In addition to the two Bt-rice strains, at least two others—one containing a bacterial blight-resistant gene, and the other herbicide-tolerant—had been under consideration at the national biosafety committee for biosafety certification.[51]

R&D on the two varieties of Bt rice granted biosafety certificates started in 1995 at Huazhong Agricultural University's National Key Laboratory of Crop Genetic Improvement, led by Zhang Qifa. While Zhang himself had not been directly involved in the research, Tu Jumin, one of his doctoral students, got significant results on Huahui 1 and Bt Shanyou 63 when he visited the International Rice Research Institute in the Philippines between 1995 and 1998.[52] Huazhong's Peng Zhongming contributed to the breeding of the seeds. The research passed MOA's appraisal in 1999, the year when Zhang's lab started the biosafety assessment process, which, in retrospect, lasted ten years and was in strict accordance with the State Council regulations and the MOA measures, as well as international safety guidelines for agricultural GMOs. Field tests came first, followed by environmental releases in 2002, and preproduction tests between 2003 and 2004. The applications for biosafety certification were submitted for the first time in 2004. The process involved the Hubei Centers for Diseases Control and Prevention; and MOA also commissioned testing at the Chinese CDC, China Agricultural University, the CAAS Institute of Plant Protection, other testing centers, and third-party organizations.[53]

Bt rice's food safety assessment alone went through three stages. The first was nutrition evaluation, testing such major nutrients as protein, fat, starch, moisture, and ash, including amino acids and fatty acid composition, micronutrients (minerals and vitamins), and anti-nutritional factors. The second stage was mainly on the toxicology of Bt rice, including a ninety-day rat-feeding study, acute toxicity tests, genetic toxicity tests, three-generation reproduction tests, chronic toxicity tests, and observed adverse effects in experimental animals. At the third stage, the focus was on Bt rice's sensitization, including protein and homology comparison of amino acid sequences of known allergens, and the target gene's protein expression in *in vitro* gastrointestinal digestive stability tests. According to Zhang Qifa, for three years after 1999, about six metric tons of Bt rice had been fed to rats for various tests.[54]

But in both 2004 and 2008, applications from Zhang's lab failed to obtain the green light on biosafety out of concerns over ecological risks and food safety.[55] China's national biosafety committee had been as precautious and prudent as their counterparts elsewhere, subjecting Bt rice to a much more rigorous safety assessment and review process and requesting more studies on its environmental and health impacts. Such a process does not just target the Bt rice developed at Zhang's lab.

The Reaction

As noted, there was a gap of two long months between the approval of these Bt strains in August and their disclosure in October 2009, and the news of their biosafety certification found its way to China after reported abroad first. As Clive James and his ISAAA are among the enemies of China's anti-GMO camp, his comments caught the attention of Greenpeace, which immediately relayed the information to the Chinese public.

Meanwhile, MOA did not disclose any information on the health and environmental studies for the Bt-rice biosafety evaluation; such information was only made public on July 18, 2014, right before the expiration of the biosafety certificates.[56] Nor did it reveal the members of the national biosafety committee who made the final recommendation. When bombarded with inquiries about Bt rice's biosafety certification, the ministry only released a brief statement saying that the decision was made "based on a five-stage rigorous process of experimental studies, pilot studies, environmental release, production safety evaluation test, and applications for biosafety certificates for commercialization that has lasted many years, as well as the assessment of NBCAGMO."[57] Therefore, the public had reasons to be outraged. Wei Rujia, the Beijing attorney, sued MOA at the Beijing No. 2 Intermediate People's Court, requesting a disclosure of scientific data related to its approved Bt rice, but the court threw the case out for technical reasons.[58]

MOA's handling of the case also left room for criticism in the media, and especially the circulation of misinformation and accusations on the internet. For example, the announcement was characterized as "covert," or described as "bashful policy released quietly," apparently borrowing from the title of a popular song, "Bashful Roses Blossom Quietly."[59] Tu Jumin, the Bt rice's inventor, acknowledged that the way in which MOA made the announcement gave the impression of "secrecy" and "diffidence." "The public should have the right to know," Tu lamented.[60] However, as explained by Peng Yufa, the five-time NBCAGMO member, MOA always notifies the applying organizations and applicants directly of the outcome of an application for biosafety evaluation, which is also the routine practice in China's administrative approval process.[61] Since 2000, MOA had been inclined to keep information on applications for field tests and environmental release of GM crops secret and had provided only very basic information on imported GM crops, such as their country of origin, purpose, and approval numbers.[62] The situation has just begun to change.

Problems Encountered in Implementing the Biosafety Regulations

In China, having stringent laws and regulations is one thing, and their strict enforcement and implementation is quite another. Thus, chronic problems of enforcement and implementation have consistently compromised the laws and regulations. This situation is attributed to a lack of resources, tension between the goals of different bureaucracies, horizontal coordination of activities among different government agencies at the same level, and vertical conflicts between central and local governments.[63] Besides these and China's sheer size, which poses challenges for compliance and risk management in monitoring literally hundreds of millions of hectares of farmland, implementation of biosafety regulations has encountered other problems that are perhaps inherent in the development of transgenic crops.

The most noticeable challenge is the illegal planting of GM crops, which seems to be widespread and lasting. As mentioned, virus-resistant tobacco was grown before it was approved for commercialization. The management of Bt-cotton varieties has been more chaotic. For example, Bt cotton was found growing in Hebei and Henan, located in the Yellow River–Huai River cotton-planting area, as well as in Xinjiang, in 1997, before it was approved for planting.[64] As research institutes and seed companies promoted their own products, Bt-cotton seeds on the market came from a variety of sources. Early in the twenty-first century, there were some fifty kinds of Bt-cotton seeds, including approximately thirty varieties using genes from Monsanto and CAAS, sown in Anhui's Wuwei county alone; in one extreme case, more than ten varieties were planted in one village. Some varieties did not even have brand names or proper labels, presumably experimental strains without assessment for biosafety.[65]

Replacement of unapproved seeds with approved ones produced inexpensively by public R&D institutes seemed to be an effective way to eradicate the problem of the proliferation of unapproved and mostly illegal Bt-cotton varieties. In 2004, MOA released a set of complementary measures for Bt cotton, which were revised in 2008, to solve the problem. On one hand, these measures made biosafety regulations more comprehensive and stringent. For example, MOA began to require that all newly developed Bt-cotton varieties obtain a biosafety certificate before being allowed to enter regional variety tests, the last stage before production, which would be permitted only after a review of data on yield and performance. MOE also stipulated that tests be conducted in its designated facilities. On the other hand, MOA sought to simplify the certification and approval process. Any Bt-cotton variety that has received a production biosafety certificate (or commercialization certificate) from one province can now directly apply for a certificate from another province, as long as the provinces are within the same cotton-growing region (China has three such regions—Yangtze River, Yellow River–Hua River, and Xinjiang). Also, if a new Bt-cotton variety is obtained from a parent variety that has already been granted a biosafety certificate, its developer can file for a biosafety certificate as well.[66]

While the next chapter will detail the role of Greenpeace in derailing China's efforts to develop GM crops and foods, suffice it to say that this international NGO has waged a more-than-a-decade-long war on Bt rice, and especially on the leader of its development, Zhang Qifa of Huazhong Agricultural University. Through a series of investigations, Greenpeace has not only revealed illegal planting of Bt rice but also further implied a connection between the varieties and Zhang Qifa, whose lab was approved for carrying out tests on the varieties for commercialization but had never been authorized to use them for purposes other than scientific experiment. Also, Greenpeace has chased unlabeled Bt rice on the market. In a word, it has held Zhang Qifa liable for negligence in monitoring.

It all began in February 2005 when Greenpeace collected twenty-five samples of rice seeds, paddy, and rice in Wuhan and its surrounding areas and sent them to GeneScan, a special laboratory for GMOs in Germany, for testing. On April 7, the test results came back, and nineteen samples showed positive for transgenic ingredients. Ten of the positive samples had been collected from wholesale and retail markets, rice-processing factories, and homes of farmers, which Greenpeace interpreted as indicating that GM rice had been planted and sold in Wuhan and elsewhere in Hubei, and probably in other provinces. Further testing showed that the GM rice was Bt Shanyou 63, one of the Bt-rice varieties developed by Huazhong Agricultural University that was still at the experimental stage and would be granted a biosafety certificate four years later. A salesperson at the university's seed sales center indicated that it had produced more than 5,000 kilograms of seeds for a seed company in 2003. This amount of seed could be used to plant about 20,000 to 25,000 mu, or 1,300 to 1,700 hectares, which in turn could produce more than 10,000 metric tons of Bt rice.[67]

At first, both MOA and the Hubei provincial agriculture bureau denied the allegations. Later, the Hubei bureau of agriculture punished three companies, including the one affiliated with Huazhong Agricultural University, for "illegally expanding seed production." The bureau then removed some 10,000 mu, or 650 hectares, of illegally planted Bt rice with compensation to farmers for their loss. After investigations, in May 2006, MOA issued the "Guidelines of Biosafety Investigation on Field-Testing of GM Crops" to strengthen the regulation on GM crops that are under testing.

As it turns out, in late 2004, *Newsweek*, an American magazine, quoted Zhang Qifa as acknowledging that "a local company got some of the GM [rice] seed and began selling it to local farmers, . . . more than 100 hectares of GM rice are being cultivated."[68] His remarks literally tipped Greenpeace, which subsequently launched its investigation into illegal planting of GM rice.

Nine years later, on May 13, 2014, GM rice was again found in Wuhan. This time, Greenpeace claimed to have bought fifteen samples randomly at a chain supermarket and a farmer's market and sent them to a Beijing-based lab, not an MOA approved one, for testing. Four samples were found to contain GM ingredients, of which three again were Bt Shanyou 63, the variety that had been issued biosafety certificates but had not obtained final approval for commercial planting. Again, Huazhong Agricultural University claimed to have strictly followed the regulations on agricultural GMOs and suggested that Greenpeace raise the issue with MOA if it had any doubt, which it did.[69] However, in a surreal twist of events, Greenpeace emerged as the pushing hand behind the discovery of the illegal planting of GM rice. In April 2014, Huazhong Agricultural University stopped several persons from stealing seeds from an experimental field at its Hainan GM-rice base. These persons later admitted their Greenpeace campaigner identities. This incident led to an MOA circular on May 4 that R&D institutions should beef up security and management.

Probably inspired or tipped by Greenpeace, China Central Television (CCTV), the state broadcaster, commissioned tests on five packets of rice bought randomly in supermarkets, of which three were found to contain a strain of GM rice that was designed to be insect resistant. Again, it was one of two varieties developed at Zhang Qifa's lab. Aired on *News Probe*, a weekly program, on July 26, 2014, the CCTV investigation concluded that GM rice had been found on sale in Hunan, Anhui, and Fujian provinces. In an interview for the program, Zhang said that it was "not impossible" for seeds to leak into the market, for which the university was not to blame. He suggested that seeds could have been taken away in 1999 when they were first introduced to MOA. He also said, "You can't say that the seeds were leaked on purpose. It's possible that seed companies have taken away the seeds and reproduced them illegally."[70]

While clearly in violation of the regulations governing the seed varieties before obtaining production approval, illegal planting of GM crops may reflect frustration on the part of scientists and farmers. Indeed, blame had been assigned to the inaction of MOA and the Chinese government, which might have forced scientists to adopt such a strategy to advance their interests. For example, amid debates over whether China should commercialize its own GM soybeans, some scientists went so far as to test their own GM-soybean seeds in the fields or to grow foreign GM-soybean seeds without undergoing prescribed assessment. Illegal planting might also reflect the fact that "farmers do not want to wait for permission to plant approved crops."[71] As a Wuhan rice farmer appearing on the CCTV program indicated, the insect-resistant rice seeds were popular not only because they resist borers but also because they save money on pesticide and labor in the amount of RMB200 to RMB300 per mu, which is quite significant and attractive for a farmer who makes RMB1,000 per mu. Such frustration also explains the widespread illegal planting of GM crops in India so that farmers could reap the benefits.[72]

But lax enforcement of biosafety regulations has implications beyond jeopardizing China's GMO research and commercialization program. Recurrence of the illegal growing and sale of GM rice has caused possible "contamination" of the Chinese rice supply, which has adversely impacted exports of Chinese rice and rice products and will continue to do so. In recent years, a growing number of Chinese rice and rice products exported to the EU had been found to contain unauthorized GM ingredients, traces of the Bt Shanyou 63 strain to be exact, the same variety found on sale in Wuhan. In November 2013, the EU required Chinese authorities to provide a report on all rice consignments before export, instead of conducting random checks upon receiving them. This makes it necessary to strictly segregate and monitor any experimentally grown GM rice in China so as not to affect exports to countries that have yet to approve the GM-rice varieties. This, as discussed, was exactly one of the rationales for the establishment of the biosafety regulatory regime in China.[73]

China's GMO Regulations in International Comparative Perspective

China has learned from both the American and European experiences in the establishment of its biosafety regulatory regime. Similar to the United

States, where the Office of Science and Technology Policy was in charge of formulating a coordinated framework on GM crops, China's science bureaucracy, SSTC, came up with the country's first regulation of R&D on GM crops. China also similarly placed primary responsibility for biosafety regulations in MOA, with other government agencies, such as environmental protection, health, and food safety, playing secondary roles.

Apparently, China also has learned from Europe. Despite establishing no special new laws, China has formulated regulations and measures that are enforceable as laws. When its biosafety regulatory regime formally took shape at the turn of the twenty-first century, one of the most significant measures was the requirement of mandatory labeling of GMOs. While the EU has set up a threshold for labeling, in China labeling is applied not only to approved imported GMOs, and products containing such imports, but also to food products containing no trace of GMOs but using GMOs in their production. It is in this sense that the Chinese GMO labeling requirement has been considered one of the most stringent in the world. China has never formally banned commercialization of GM crops, but in actuality the period since 1997, when Bt cotton was approved for commercial cultivation, has witnessed a de facto moratorium. On top of China's biosafety regulations are new institutions established specially for the new process, such as the Joint Ministerial Conference, which decides important issues and coordinates efforts in different government agencies on GMOs, and the National Biosafety Committee on Agricultural GMOs, which is responsible for approving biosafety evaluation from research to commercialization.

Meanwhile, like India, China has faced the challenge of regulating GM crops and foods. Both countries have recently been in regulatory limbo on approval of commercializing new GM crops—Bt rice and phytase maize in China and Bt brinjal in India. China has witnessed the domination of domestically developed Bt-cotton seeds, which may have had something to do with protectionism, in contrast to India, where the Bt-cotton market has been monopolized by Monsanto genes. China and India also differ in their respective leading regulators of biosafety. Presumably, a country is oriented toward precaution if its biosafety regulator resides within an environmental protection apparatus, which, given its mission, is likely to pay more attention to the potential risks that GM crops might have for the environment. This is the case in India, whose government has been particularly consistent and adverse in handling biosafety regulation. In China, however, biosafety regulation of GMOs is under the agricultural ministry's jurisdiction. Although

Bt cotton has been the only GM crop approved for commercial planting on a large scale, other GM crops have also gone through various stages of biosafety assessment, and most significantly, Bt rice and phytase maize were issued biosafety certificates for commercialization in 2009 (these biosafety certificates were allowed to expire before being renewed).

China's GMO policy has been as paradoxical as those adopted in the EU and India. This is especially the case as China's policy allows the importation but not the cultivation of GM soybeans and corn. Since 1997, China has imported increasing amounts of GM soybeans not only because imported soybeans are cheap and high in quality, but also because indigenous ones cannot meet the rising demand for edible oil and animal feed. It also makes more economic sense for Chinese farmers to grow corn than soybeans. Therefore, annual imports have recently been over 50 million metric tons—52.6 million in 2011, 57.5 million in 2012, 63.4 million in 2013, 71.4 million in 2014, and more than 80 million in 2015, 2016, and 2017. But edible oil made from GM soybeans has to be labeled, and it took MOA three years to issue biosafety certificates for imports of three GM-soybean events developed by Monsanto and BASF in June 2013, even though these events—herbicide-resistant soybean CV127, insect-resistant soybean MON87701, and herbicide-resistant and insect-resistant soybean MON87701 × MON89788—had been approved in the United States, Canada, Brazil, Argentina, Japan, and even the EU for either commercial planting or consumption. Not only are domestically developed GM soybeans not ready to meet the challenge of imports; in fact, China may have decided to give up efforts to commercialize GM soybeans.[74]

Summary and Discussion

It is clear from this discussion that China has established a relatively rigorous and comprehensive biosafety regulatory regime to assess risks of GM crops and to achieve effective safeguards for their use. The introduction of the State Council regulations and MOA measures became a watershed in the development of GMOs in China. While China's regulations on GMOs were perceived to fall between those of the United States and Europe,[75] the government has given increasing attention to regulations on biosafety, reflected in the stipulation that 17 percent of the funds for the Mega-Engineering Program for new GMO varieties be earmarked for risk assessment and biosafety evaluation, which are extremely costly and time-consuming. The program also called for establishing a large number of independent third-party evaluation centers and institutions to collect more scientific data for science- and evidence-based policymaking.[76]

The examination of biosafety certification of Bt rice and its aftermath further indicates a delicate position taken by MOA, and indeed, the Chinese government, toward the commercialization of GM crops. While transgenic technology presents unprecedented opportunities to ensure food security, its possible risks to human health and the environment have cast a shadow on the full utilization of this modern technology.[77] Such a dilemma has caused tensions not only between scientists and other stakeholders in advancing their respective interests, but also between different bureaucracies in fulfilling their respective missions. On one hand, Chinese agricultural biotechnologists have the incentive to promote their research for obvious professional reasons, if not the commercial interests that some of them may also have. They have been vocally complaining that the current biosafety regulatory regime is too precautionary or even preventive to develop transgenic technology.[78] They have become frustrated and felt under siege, not only because of international pressures, mentioned in this chapter, and domestic resistance, to be discussed further, but also because the government has gradually chosen a precautionary policy. Consequently, although China began research on agricultural GMOs at roughly the same time as its international counterparts, the research has long remained in the labs and out of the daily lives of Chinese consumers. The government has invested substantial funds in developing GM crops, but the outcome has lagged far behind, so pressure is mounting over the commercialization of GM crops. On the other hand, while the Joint Ministerial Conference mechanism has been in place to coordinate efforts of different ministries, biosafety is largely falling within the agricultural ministry's jurisdiction, which, as opposed to MOST, has been more hesitant toward commercialization of GM crops. Surely, the environment ministry's position can be more precautionary if not preventive.

But China's biosafety regulatory regime is weak because the State Council regulations and the MOA measures are not laws within China's legislation framework. Moreover, the State Council regulations have been perceived as merely a "hat" put on a ministerial-level administrative regulation.[79] These regulations and measures are limited to the administration of technical aspects of GMOs. The biosafety regulatory regime has also not set clear stipulations as to who has the authority and how to monitor those involved in the implementation of the regulations and measures. In the case of GM food safety, while the health commission is responsible, there has been no specific legislation other than regulations from multiple government agencies. The regulations and measures also tend to be ad hoc, not comprehensive and systematic. Therefore, China needs to enact an over-arching biosafety law to guide the increasingly complicated issue of bio-safety, relating to research, commercialization, trade, food safety, and other aspects of GMOs. There have been bureaucratic competition and conflicts as the drafting of the biosafety law was assigned to the environment ministry, whose relationship with other ministries has led to a stalemate.[80]

The implementation of biosafety regulations concerning GM crops has been challenging as the lack of control over these crops in field tests has led to the introduction of untested seeds and sales of unapproved products. As well as defying existing regulations, incidents of illegal growing and sales of GM crops have not helped the research on and commercialization of GM crops, including Bt rice. To a certain extent, it is the scientists themselves who are to blame for the shift in GMO policy from promotional toward precautionary.

There is no doubt that China's evolving policy toward GM crops, as evident primarily in the case of the Bt-rice biosafety certification, has taken its cue from international development. Having closely monitored the global debates on GM crops and foods, the Chinese government has gradually but firmly taken extremely careful steps in approving new GM crops, allowing commercialization only after extended tests indicating the absence of any risk and thus considerably slowing the required biosafety certification process. The initial establishment of the biosafety regulatory regime and its enhancement, as well as the government's hesitation after 1997, when Bt cotton was approved for commercial planting, and after 2009, when Bt rice was certified biologically safe, can be attributed at least partially to such strong external influences. It is further speculated that the actual reason for inaction on the Bt-rice commercialization may well be that the United States has deregulated GM rice but has never seen it cultivated.[81]

Finally, the evolving story around Bt rice is extraordinary in that it is not for lack of endeavor by scientists on research and commercialization. The fact that the government allowed the biosafety certificates to lapse has raised the question of why it should have given the go-ahead in the first place. Of course, MOA and the central government may not have foreseen the kind of pressure and challenges that it would have to endure.[82] This time, the relatively simple part of the "science for policy" has been overtaken by the complexities of the "policy for science."[83] The government had to reconsider its initial position, factoring in other factors, especially the public's objection to GM crops and foods, in the policymaking process. Inevitably, it was a setback for scientists, who have been privileged over the last three decades. Therefore, both scientists and government officials are frustrated and upset.

CHAPTER V

Polarization and Politicization of Transgenic Technology

While China's establishment of a biosafety regulatory regime in the late 1990s was mainly influenced by external developments, nascent domestic GMO activism, which started to move to the forefront at the turn of the twenty-first century, has followed the trend of an increasingly unfriendly and even hostile environment toward GM crops in Europe and the global attention given to biosafety and biodiversity. Indeed, global attitudes toward GMOs have polarized between proponents who advocate that GMOs have been proven safe and environmentally benign, reducing pesticide use while benefiting farmers, and opponents who stress the uncertainties and possible environmental and health risks associated with GMOs and question the objectivity of scientists and corporations in developing GMOs.[1] The global debates have also become fervently politicized, linking the commercialization of GM crops in developing countries with a deliberate American plot to threaten food security, food sovereignty, and even national security in these countries.

This chapter examines a similar and escalating feud between the research community and the public over GM crops and foods in China. Along with the debates on GM crops and foods in general, it will give particular attention to how the issuance of biosafety certificates to Bt rice, a staple crop, and the trial of Golden Rice among Chinese children have been politicized, which in turn has shaped public perception of the research and commercialization of GMOs. The role of Greenpeace, which has been increasingly prominent in fighting against the introduction of GM crops, and especially GM rice, in China and elsewhere, will also be analyzed in this context. Finally, the chapter will discuss the rise of a rampant "conspiracy theory" that has maliciously framed transgenic technology and Chinese agricultural biotechnologists. Together, these developments may have helped shift China's GMO policy toward a precautionary position.

The "Right to Know"

At the turn of the twenty-first century, as noted in the last chapter, the Chinese government started to formulate regulations and measures to better manage the R&D and commercialization of GM crops. China also became a net importer of soybeans, which were increasingly products of genetic modification and in turn were likely used to make edible oils and tofu for domestic consumption. This backdrop stimulated a campaign for the "right to know" (*zhiqingquan*).

The "right to know" can have several meanings. Superficially, consumers want to know whether their foods contain GM ingredients. But first and foremost, consumers need facts and evidence on GM foods and their potential impacts on health and the environment to make an informed decision. Without such information, consumer's "right to know" would be incomplete and partial. Written into the Chinese regulations, the mandatory labeling requirement seemed to signal that GM foods are different and by implication unhealthy, and consumers were granted the "right to know" conditional upon the labels but not necessarily the knowledge of the effects of GM foods.

Around that time, Greenpeace, the international environmental NGO, operating out of its Hong Kong office, started to recruit mainland Chinese college students as volunteers to purchase dairy products in supermarkets and send them to Hong Kong for testing for the presence of GM content. On December 6, 2002, Greenpeace released a list of GM-ingredient-containing products sold in China, including those made by Nestlé, the Swiss food company, that were not labeled. According to Sze Pang Cheung, Greenpeace's China GM campaign manager, "We are talking here about a company which has repeatedly been found to be selling GM products without informing consumers, and have vowed to continue using GM ingredients in China."[2] One of the Nestlé products targeted by Greenpeace was Nesquik, an infant formula.

Encouraged by Greenpeace, in March 2003, Zhu Yanling, a Shanghai resident, sued Nestlé for not labeling Nesquik, which she had been buying for her three-year-old child, thus spearheading a campaign against GM foods in China. A court in Shanghai first commissioned the Center for Biotechnology under the Shanghai Academy of Agricultural Sciences to test the product, which was found to contain ingredients derived from GM soybeans. Nestlé rejected the test results on the ground that the testing used a methodology not approved by the Ministry of Agriculture. The court then ordered a second test to verify the findings, which turned out negative for the presence of GMOs, and rested the case accordingly.

But the story did not end there. In December 2003, Greenpeace further supported Zhu to directly confront Nestlé's top management by flying her and her lawyer to Switzerland. Zhu expressed outrage that the company did not believe Chinese consumers were as conscious about GM foods and consumer rights as Europeans and that the company would continue using GM ingredients in its products sold in China. To her, Nestlé's position represented a double standard that, though not necessarily legally wrong, was morally and ethically unacceptable. She therefore demanded that Nestlé adopt the same policy in China as it did in Europe and called on the company to respect consumers' right to an informed choice by properly labeling its GM products and phasing out GM ingredients. Nestlé rejected the demands.

While the media coverage of the case, to be discussed in next chapter, made some Chinese more suspicious about Nestlé and GM foods, Greenpeace further pressured Nestlé by urging Chinese to cast their votes on Sina.com, one of China's most popular internet portals, to rally behind Zhu and to denounce Nestlé's stand on GMOs in China. Later, Greenpeace itself released results from an unrepresentative survey of 600 consumers in Beijing, Shanghai, and Guangzhou on the issue, claiming that 87 percent of the respondents demanded labeling GM products and 70 percent of them would lose confidence in a brand if its products were found to contain GM ingredients.[3] As Greenpeace whipped up anti-GMO sentiment, some Chinese consumers consciously followed up their online action with offline action by returning Nestlé's products for refunds and boycotting them whenever a choice was available. In 2004, Greenpeace expanded its list of food products sold in China to be avoided because of the presence of GM ingredients. The "right to know" has since become the Chinese public's most powerful weapon to insist upon strict labeling of GM foods in general and edible oils crushed from imported GM soybeans in particular, as such oils now dominate China with a 90 percent market share.[4]

Opposition to the Commercialization of GM Rice

Over the next decade, the debates over the pros and cons of the development of GM crops and especially the introduction of GM foods intensified. Both sides of the GMO divide seem unable to convince the other. From time to time, both have used impolite and even insulting language to attack the other. The debates further heated up in 2009 when the news was made public that biosafety certificates for commercialization had been issued to Bt rice and phytase maize.

On February 5, 2010, some fifty members of Huazhong Agricultural University's National Key Lab of Crop Genetic Improvement, led by Zhang Qifa, gathered to celebrate the forthcoming Chinese New Year. The foods on the table—sushi, porridge with sweet potato, and *doupi*, a local treat—were all made with the Bt rice developed at the lab.[5] The lab had reasons to celebrate as the past several years had meant so much for it, and indeed, for Chinese transgenic scientists. In 2008, the State Council launched the GMO Mega-Engineering Program, promising to channel some RMB20 billion into research and commercialization of new transgenic varieties. In 2009, two strains of Bt rice, which had been developed at the lab for more than a decade, were finally granted biosafety certificates for commercialization. About a week earlier, on January 31, Central Document No. 1 clearly accentuated the thrust of implementing the GMO MEP based on regulations and scientific assessment. It seemed that Bt rice would soon appear on Chinese dining tables.

Scientists also increased their media appearances. They explained the procedure that the Bt-rice varieties had gone through before obtaining biosafety certificates; they provided details of the food-safety evaluation process, including who did what, how it was done, what the outcomes were, and what they meant; they promulgated the agronomic, health, and environmental advantages and benefits of Bt rice, and GM crops in general; and they answered questions such as whether consumers should eat unnatural foods such as GM foods, and especially whether they should consume Bt rice that insects do not eat. Through interviews and writing, Fang Zhouzi, the U.S.-trained biochemist turned science writer, also joined Chinese biotechnologists in their endeavor to dispel the negative image of transgenic technology and GMOs. In a word, members of the Chinese GM research community anticipated and attempted to preempt the public's concerns and attacks from the GMO activists.

Meanwhile, the Chinese public heightened their concerns over the safety of Bt rice, which was not unexpected given the central role of rice in their diet. They thought that biosafety certification would automatically lead to commercial planting, without realizing that before being put on their dining table, the Bt-rice varieties would still need to undergo value for cultivation and use (VCU) verification, variety registration, and seed-licensing procedures, during which the varieties would be subject to further testing and examination and might be found unsuitable for registration, rejected for seed-license application, and/or turned down by the market because of commercial infeasibility.

The certification process and the announcement of the issuing of biosafety certificates for Bt rice and maize, which (as alluded to in chapter 4) were not as transparent as they were supposed to be, sowed more doubts about the situation. Information on the national biosafety committee's membership and the basis on which Bt-rice varieties had been certified biologically safe were also not made available on a timely manner. In an environment that was unfriendly, biased against, and hostile toward GM crops and foods, which will be further examined, some Chinese also perceived the National Biosafety Committee of Agricultural GMOs to be compromised because most of its members have business interests in the technology under review.

These concerns should be viewed in context. Given the recent outbreak of SARS and a series of food-safety scandals—from melamine-contaminated baby formula to meat products containing clenbuterol, a chemical that is hazardous to humans—Chinese consumers thought that GM foods could be the next hidden food-safety danger. Bt rice found on the domestic market and in rice exports, even though commercial cultivation had never been authorized, pointed to severe lapses in the governance of food safety and biosafety and the limited capacity of the Chinese party-state.[6] Thus, the public's outcry over the biosafety certification of Bt rice reflected an overall lack of trust in government's role in regulations, and especially in the implementation and enforcement of the regulations.[7]

The anti-GMO camp was mobilized against the developments that transgenic scientists had cheered for, amid the public's rising concerns. Before the annual sessions of the National People's Congress and the Chinese People's Political Consultative Conference in March 2010—led by Su Tieshan, a former deputy secretary-general of the Society of the History of the People's Republic of China; Zhang Hongliang, a professor at the Central University of Nationalities; and Li Changping, a well-known advocate for farmers' rights at Hebei University—some 120 scholars petitioned NPC opposing the commercialization of Bt rice in China, setting the stage for an anti-GMO showdown. Su, Zhang, and some petitioners were associated with Utopia, an organization of the Chinese "New Left" with a website of the same name, known for extolling the late Chinese leader Mao Zedong, damning capitalism, and attacking globalization.[8] The petition especially stressed:

> Against the global trend of commercialization of bio-energy technology and weaponization of agricultural products, in November 2009, the Ministry of Agriculture approved the commercial planting of transgenic rice and maize. China will become the first country in the world to genetically modify a staple crop (GM crops in the other countries are limited to non-staple crops). We believe that genetic modification of a staple crop when concerns over its safety remain widespread and intensely debated is likely to endanger the Chinese nation and China's national security.

The petitioners urged the State Council to withdraw the biosafety certificates issued to Bt rice and maize, immediately halt the commercial production of GM staple crops, and impose strict administration over the production of GM foods. They further requested that the state draft a law on the research and application of GMOs and ban all government agencies from initiating new commercialization involving GMOs. According to the petition, the legislation should ensure that R&D, experimentation, promotion, and application of GM seeds be state-led, similar to the legal requirements for nuclear technology; that producers notify the public of any trial and experimentation with GM seeds; and that consumers be notified of the entry of GM foods into the market. Lü Xinyu, one of the petitioners and a professor at Fudan University's School of Journalism, stated sensationally that the petitioners did not bother to debate with genetic engineers because these so-called experts were not trustworthy and had vested interests in developing GMOs. "We can use our common sense to question the safety of GM foods," said Lü. Su Tieshan, one of the leading petitioners, also indicated that the commercialization of transgenic rice should not be decided by a small number of scientists and experts; instead, the matter should be discussed by all Chinese and subject to a law enacted by NPC.[9]

CPPCC, which also was in session, received two GMO-related proposals from several dozen of its members—one led by Dong Lianghui, a daughter of Dong Biwu, and the other by Ren Yuanzheng, a daughter of Ren Bishi (both senior Dong and senior Ren were among the founders of the People's Republic). Both proposals supported China's concentrated and concerted efforts to develop transgenic technology so as to become a global leader but opposed the commercialization of Bt rice without ensuring its safety for consumers.[10]

Around that time, a group of scholars of science, technology, and society, the history and philosophy of science, ethics, and other disciplines in the social sciences and humanities posted an open letter calling for postponing the commercialization of Bt rice. They claimed that China does not own intellectual property rights (IPRs) to the Bt-rice varieties to be commercialized; there is no evidence that GM crops increase yields; GM crops potentially pose risks to human health and the ecological environment; and the public should have the right to know about the consequences of introducing GMOs as staple foods that are fundamental to people's livelihood and future generations.[11] This was a more reasonable appeal. As it was posted on ScienceNet, a website administered by the Chinese Academy of Sciences, the Chinese Academy of Engineering, the National Natural Science Foundation of China, and the China Association for Science and Technology, the appeal attracted quite a number of cosigners among ScienceNet's bloggers.

On November 26, 2010, an extraordinary exchange broke out between pro- and anti-GMO camps during Zhang Qifa's lecture at China Agricultural University in Beijing. The lecture was irrelevant to GMOs, but Zhang was asked about the safety of GM crops and especially whether scientists had illegally distributed transgenic seeds to farmers. Zhang and the host refused to entertain the questions, insisting that a public forum would be better suited to address such concerns. Suddenly, a middle-aged woman stood up and shouted: "You are a traitor! You are using 1.3-billion Chinese as guinea pigs to serve your American master!" Then, a person identifying himself as an engineer rushed to the front of the podium, trying to snatch the microphone, while several others upheld a banner that read "Objection to the commercialization of GM rice!" When approaching the podium, an older man became so emotional that he picked up a ceramic tea mug from the front row and hurled it, missing Zhang Qifa.[12] This was not activism, but thuggery, and likely the first personal attack on Chinese scientists since the end of the Cultural Revolution in 1976.

The raw, lingering emotion between the pro- and anti-GMO camps continued to erupt vehemently from time to time. On January 11, 2012, for example, when both sides were tape-recording a program at Beijing Television, a number of participants in the audience, presumably from Utopia, waved their arms and cursed the others as "Traitors! Liars!" Some threw their boards to the floor. Fang Zhouzi, the science writer and one of the guests, furiously pounded the table, demanding that those disturbing the program be removed.[13]

Greenpeace in China

Having become heavily involved in GMO activism in China, Greenpeace is undoubtedly the most influential but also an unlikely player in China's domestic affairs, given its past encounters with the Chinese government. In the early morning of August 15, 1995, at Beijing's Tiananmen Square, six Greenpeace activists displayed a banner showing "Stop All Nuclear Testing!" in both English and Chinese. The following June, Greenpeace sailed one of its vessels, *Greenpeace*, into China's territorial waters and dropped anchor at the mouth of Shanghai harbor, as well as rallied in Barcelona, Spain, during the visit of Chinese president Jiang Zemin, to protest China's underground nuclear tests in Lop Nor. In 1997, Greenpeace returned to China, setting up an office in Hong Kong. In 2001, it established a campaign office in Beijing, which in 2003 filed a corporate registration application under the name of Beijing Rainbow Peace Environment Research Center, not only keeping a low profile but also concealing its real identity.[14] Now, Greenpeace's operation in Beijing has a staff of sixty to seventy and cooperates with some Chinese government agencies, research institutes, and environmental NGOs on issues of food safety, imports of electronic waste, and climate change. As China was turned into a new battleground over GMOs, Greenpeace has become the major activist organization behind the operation of opposing the introduction of GM crops and foods, locked in a battle with agricultural biotechnologists.[15]

In March 2002, Greenpeace started to collaborate with the Nanjing Institute of Environmental Sciences under the State Environmental Protection

Administration, predecessor of the Ministry of Environmental Protection and now the Ministry of Ecology and Environment. Through joint publication of *International Biosafety Newsletter* and *Biosafety Translation Series*, they disseminated mainly negative information on GMOs. In early June 2002, under the name of the Nanjing Institute, Greenpeace released a major report, "A Summary of Research on the Environmental Impacts of Bt Cotton in China," with Xue Dayuan, the expert on biodiversity at the Nanjing Institute, as the author. As well as concluding that Bt-cotton varieties would be ineffective in controlling bollworm after eight to ten years of continuous production, the report indicated that many Bt-cotton farmers continued using pesticides; that the susceptibility of bollworm to the Bt toxin fell to 30 percent after seventeen generations under continuous feeding with Bt-cotton leaves, while the resistance of the bollworm increased a thousand times when the feeding was continued to the fortieth generation; and that secondary insects had replaced the bollworm as the main threat to Bt cotton in certain areas.[16] China's leading media, including *China Daily*, the English-language mouthpiece of the country's ruling Chinese Communist Party, and the official Xinhua News Agency covered the report and the seminar held on the occasion of the report's release. However, not only ignoring the large economic and health benefits that small farmers received from growing Bt cotton, the report also selectively used unpublished research by Wu Kongming, an expert on plant protection at the Chinese Academy of Agricultural Sciences; in fact, the report contradicted a series of papers that Wu would soon publish.[17]

But this early approach of framing GMOs as a biosafety or biodiversity concern for causing the spread of secondary insects and environmental contamination did not resonate well among the Chinese public, who were unable to directly relate biosafety or biodiversity to their daily lives. Greenpeace then tactically shifted its focus toward food safety associated with the GM foods.[18] Zhu Yanling's case against Nestlé, described above, constituted such a change, encouraged and supported by Greenpeace in the name of defending the "right to know" and concerns about food safety. More than rhetorically opposing GM crops and foods, Greenpeace executed a sophisticated strategy using Chinese media to effectively disseminate its campaign messages.[19] It became more provocative in its campaign against GMOs in China when approval for the commercialization of GM rice seemed imminent.

In 2004, Greenpeace spread news that China soon would allow the planting of GM rice, flaring tensions between scientists and the public.[20]

It alleged that multinational companies such as Monsanto, rather than Chinese farmers and consumers, would stand to profit from the commercialization. It claimed that foreign companies, rather than domestic scientists, owned the patents used in the development of Chinese GM rice (see further discussion in chapter 7). And it engaged in the search for the illegal planting and sales of GM rice. As discussed in chapter 4, in 2005, Greenpeace claimed to have found Bt gene in rice sold in Hubei; between July 2009 and February 2010, it further alleged that rice and rice products sold in Hubei, Hunan, Fujian, and Guangdong provinces contained Bt gene; and in the following months, Greenpeace claimed to have found rice with GM ingredients in farmer's markets in Xiantao, Hubei Province, and Changde, Hunan Province. While attributing some of the GM ingredients to Huazhong Agricultural University's Bt Shanyou 63, Fang Lifeng, a Greenpeace China campaigner, also alleged that GM ingredients might have come from other domestic labs or even foreign sources. After Bt rice finally received biosafety certificates in August 2009, Greenpeace intensified its campaign. It continued its mission of ferreting out the illegal cultivation and distribution of GM rice as recently as May 2014. Such intensive campaign efforts generated an impression that GM-rice contamination was intentional and widespread.[21]

In a word, Greenpeace had launched an all-around assault on GM rice on the fronts of the environment, IPRs, and illegal cultivation and sales, in addition to its embrace of the "right to know" and choose and the labeling of GM foods. It stirred up unprecedented attention to GM foods, and Bt rice in particular, in China, mobilizing the public to voice their concerns and eventually placing the issue on the agenda of the political leadership. In fact, Greenpeace's strategy seemed effective. For example, Greenpeace itself attributed the postponement of the biosafety certification process for Bt rice in 2004 and 2008 to its campaigns;[22] and it also attributed the call for the development of indigenous IPRs in 2010's Central Document No. 1 to its efforts, although at the turn of the twenty-first century China started to emphasize the generation of indigenous IPRs as the core of its overall innovation strategy.[23]

But a twist put Greenpeace on the hotspot. At around 9:40 PM on April 11, 2014, three members of Greenpeace China were apprehended for an attempted theft of GM-rice seeds and leaves from an experimental field at Huazhong Agricultural University's Hainan research base. These campaigners claimed to be on an "undercover field investigation" mission to find loopholes in the safety management of GM rice. After an illicit cultivation of GM corn in Hainan Province was reported in March 2014, explained Wang Jing, director of Greenpeace China's food and agriculture program, her organization wanted to confirm that such illegal activities came from Huazhong Agricultural University, which had been conducting the experiments in Hainan, and to monitor whether GM rice had been illegally released into the environment. But Wang also acknowledged that her campaigners had found no evidence of the spread of GM rice before deciding to intrude into Huazhong Agricultural University's research base.[24] The incident raised suspicion whether Greenpeace had gotten the rice sample with GM ingredients from experiment sites rather than the market, which would completely undermine and discredit its campaign.

The Golden Rice Controversy

Greenpeace was also behind another campaign alleging that American and Chinese scientists had tested Golden Rice, a type of rice genetically enriched with β-carotene, a precursor of vitamin A, in a group of Chinese children without proper research ethics clearance, informed consent, and follow-up monitoring. On August 1, 2012, the *American Journal of Clinical Nutrition* published a paper online showing that Golden Rice is a promising source of vitamin A in Chinese children ages six to eight. On August 31, spotting the paper, Greenpeace immediately condemned the study for using Chinese children as "guinea pigs," calling it a "scandal of international proportions."[25]

The lead author of the study was Guangwen Tang, an associate professor of nutrition of Chinese origin at Tufts University in the United States. In addition to three U.S.-based coauthors, the study had three Chinese collaborators—Hu Yuming of Hunan Province's Centers for Disease Control and Prevention, Yin Shi-an of Chinese CDC's National Institute for Nutrition and Food Safety, and Wang Ying of the Zhejiang Academy of Medical Sciences.[26] The study was funded by the U.S. Department of Agriculture, the National Institute of Diabetes and Digestive and Kidney Diseases under the National Institutes of Health, as well as China's National Key Technology R&D Program, also known as the *Zhicheng* Program, under the Eleventh Five-Year Plan for S&T Development (2006–2010). While a partially American-sponsored study involving a type of GM rice was already sensitive enough, the experiment through which scientists obtained the outcomes was more problematic.

Indeed, there were several lapses or even serious violations surrounding the study. First, according to the regulations governing research on GMOs in China, as discussed in chapter 4, approval from the national biosafety committee is required for carrying out a study even if the GM material used has been proved biologically safe outside China. Tang originally sought approval from the Zhejiang Provincial Agriculture Bureau through the Zhejiang Academy of Medical Sciences where she held a visiting appointment. On July 4, 2008, Greenpeace tipped off MOA's Office of Agricultural Genetic Engineering Biosafety Administration, which in turn ordered the Zhejiang agricultural bureau to investigate the matter. As the research team had not applied for approval to import Golden Rice or to conduct the experiment, the Zhejiang authority ordered researchers to halt the experiment, effective immediately;[27] but the twenty-one-day experiment, which started on June 2, 2008, had already ended.

Second, on the other side of the Pacific, the U.S. federal government also requires that all research involving human subjects undergo an institutional research ethics review, let alone one carried out in another country where six- to eight-year-old children were administered with Golden Rice. According to the published paper, the study's recruitment processes and protocol were approved by "the Institutional Review Board–Tufts Medical Center in the United States and by the Ethics Review Committee of Zhejiang Academy of Medical Sciences in China," and "both parents and pupils consented to participate in the study." In reality, information available at the U.S. government's clinical trial registration website indicated that the trial would take place in the United States, without mentioning China; that is, there might have been an intention to hide the nature and location of the study from the very beginning. The experiment was eventually carried out in Hunan province, not in Zhejiang where Tang and her team had initially intended to seek approval. Before the experiment began, the team did meet with the students and their parents, informing them that the study was about nutrition collocation and vitamin A absorption; but the consent form only mentioned β-carotene-containing rice instead of Golden Rice or GM rice. The chain of events suggests that researchers did not follow ethical protocols for human research, that they broke quite a few laws and regulations, and that there were serious lapses in judgment on the part of the researchers themselves if not the American and Chinese regulators.

Both the Chinese government and Tufts University immediately launched investigations. The three Chinese collaborators initially tried to distance themselves from the study: Hu denied that Golden Rice had been used, Yin indicated that he was unsure whether Golden Rice had been used, while Wang claimed that she did not know about the paper at all. But a three-month investigation, led by the Chinese CDC, rendered a decision, on December 6, 2012, to sack the three Chinese coauthors of the study. Parents of the children fed with or without Golden Rice received compensation of RMB80,000 ($13,000) and 10,000 ($1,600), respectively, from the government. The National Health and Family Planning Commission also released new guidelines on research involving human subjects to safeguard the interests of those participating in clinical trials, whose number has been increasing.[28]

In September 2013, Tufts University completed its own investigation, confirming that Guangwen Tang had violated regulations on research ethics while studying Chinese children using Golden Rice. Specifically, the university characterized Tang's activities as constituting serious and continuing noncompliance with federal regulations and with internal human-subject review policy at Tufts. In particular, Tang did not adequately explain the nature of Golden Rice in the informed-consent process; she did not get further authorization for changes in the study protocol from Tufts' institutional review board responsible for reviewing all research involving human subjects; and she did not describe the participation of Chinese researchers in the protocol. Consequently, Tufts suspended Tang for two years from conducting human research, during which time she would be "retrained on human subjects research regulations and policies"; after the training was completed, for a further two years she would be allowed to conduct human studies only as a supervised coinvestigator. The external reviewers also criticized Tufts for having failed to verify whether research ethics panels in China were capable of reviewing the study and whether the Chinese panels actually had reviewed and approved it.[29] On July 29, 2015, the *American Journal of Clinical Nutrition* retracted Tang's paper, after Tang failed to block the journal's publisher, the American Society for Nutrition, from doing so at the Massachusetts Superior Court, claiming that the retraction would constitute defamation.[30]

There is no doubt that Guangwen Tang bypassed Chinese regulations regarding research on GMOs by transporting Golden Rice into China illegally. Tang and her Chinese collaborators also fabricated research ethics approvals by relevant American and Chinese institutions, and violated regulations governing research on human subjects without properly seeking approval for and following study protocols while performing the experiments on children in China. However, both internal and external reviews convened by Tufts found no evidence of health or safety problems in the children fed with Golden Rice; they also concluded that the study's data were scientifically accurate and valid, and that the results "have important public health and nutrition implications, for China and other parts of the world."[31] This was good news for an episode that otherwise was faulty.

However, this case of violation of regulations on research ethics was played up to attack the conduct of research on GMOs in China as a whole. Facing strong and widespread suspicion of GMOs, which may or may not be rational or scientific, Chinese researchers found it almost impossible to find a group of parents in any classroom in China who would agree to have their children participate in an experiment involving Golden Rice. This explains why during the study a Chinese official emailed Guangwen Tang indicating that he was taking any mention of transgenes (*zhuanjiyin*) off the informed consent, because the term was just "too sensitive." There is also speculation that Tang's Chinese collaborators were removed from their positions simply to ease the anti-GMO backlash in China.[32] Nevertheless, the Golden Rice controversy has seriously eroded "public trust and tainted the reputation of research on genetically modified crops," said Lu Baorong, a Chinese ecologist and a five-time member of NBCAGMO.[33]

"Conspiracy Theory"

One of the strategies used by GMO activists in China is to associate transgenic scientists not only with their own professional and business interests but also with an American conspiracy to undermine China. The stakes in the Bt-rice debate have thus been escalated to a level that would have political implications for food security and sovereignty, national security, and even survival of the Chinese nation. In a word, GMOs are the embodiment of evil.

Conspiracy theory about GMOs is not a Chinese invention or a China-specific phenomenon. It may have originated from Jeffrey M. Smith, an American consumer activist and self-published author, in his books *Seeds of Deception: Exposing Industry and Government Lies About the Safety of the Genetically Engineered Foods You're Eating* and *Genetic Roulette: The Documented Health Risks of Genetically Engineered Foods*;[34] or F. William Engdahl, a German-American, in his book *Seeds of Destruction: The Hidden Agenda of*

Genetic Manipulation;[35] or Marie-Monique Robin, a French TV journalist and documentary filmmaker, in her book *The World According to Monsanto: Pollution, Corruption, and the Control of Our Food Supply* and her documentary with the same title;[36] or a combination of all these and probably others (most of the books mentioned here have been translated into Chinese and published in China). In particular, Engdahl, a self-proclaimed "award-winning geopolitical analyst, strategic risk consultant, author, professor and lecturer," according to his personal website,[37] has received extremely rare treatment in certain quarters of Chinese society for a foreigner. Despite not being published by an academic or a reputable publisher, his 2007 book on genetic manipulation was immediately translated into Chinese with *Food Crisis* as its title.[38] Engdahl has also been very popular in China's speaking circuit and blogosphere.[39]

In his book on genetic manipulation, as well as his speeches and interviews given in China, Engdahl has advanced a theory about "evil" transgenic technology. His arguments are as follows. First, transgenic technology does not represent progress in science and technology but is a political plan to use human beings as "guinea pigs." Second, transgenic technology is just a reprise of eugenics, the belief and practice of improving the genetic quality of the human population, which became an ethnic extermination experiment in Hitler's Nazi Germany in the 1920s and 1930s, supported by the Rockefeller Foundation. Third, transgenic technology is part of the U.S. elite's attempt to control the global food chain and to especially target populous countries like China and India. Fourth, proven to be scientifically dangerous, transgenic technology brings incalculable risks to life on the Earth. And fifth, the promotion and proliferation of GM crops, especially in Asia, ultimately aims to launch a vicious new biological war. Engdahl concluded one of his speeches in China this way:

> Now we already know that the Pentagon has been engaged in extremely secret research to improve the nano-bio-cognitive weapons. These weapons utilize genetic engineering and nanotechnology to target specific ethnic groups. With the perfection of these technologies and their incorporation into the widespread GMOs in Asia, Rockefeller Foundation would make its eugenic dream a reality by wiping out billions of people of color from Earth. GMOs would launch a newer, broader, and more frightening biochemical version of the "Opium War," in which civilization may not survive.[40]

Indeed, Engdahl's book and speeches contain all the ingredients of a "conspiracy theory." Backed by the Rockefeller Foundation, Monsanto, a notorious American MNC that has been a punchline as far back as the Pentagon's secret biological warfare program and the Vietnam War, when dioxin and Agent Orange were used, has attempted to use GM crops to control the global food chain and to annihilate Chinese. In particular, according to Engdahl, generous grants from the Rockefeller and Ford foundations brought leading Third World agricultural scientists and agronomists to the United States for training in modern agricultural business so that they would then bring back the knowledge to their homeland. In the process, "they created an invaluable network of influence for US agribusiness in those countries, all in the name of science and efficient free market agriculture."[41]

One of China's leading anti-GMO personalities is Gu Xiulin, a U.S.-trained agricultural and resources economist. She not only translated several of Engdahl's books into Chinese but also published a book, *GMO War: Defending China's Food Security in the Twenty-First Century*, reiterating many of the allegations in Engdahl's *Seeds of Destruction*.[42] For Gu, transgenic technology is fundamentally "evil," aiming to wipe out human beings, and its diffusion has an unimaginable consequence of biological destruction. She has taken an uncompromising and intolerant attitude toward transgenic technology and GMOs:

> I am fully and completely opposed to the application of transgenic technology on agriculture, not one application! Neither human nor animals should consume GMOs! GMOs should not be grown and sold. Never, ever![43]

Along the same line, Tong Pingya, a retired corn researcher at the CAAS Institute of Crop Sciences, charged that since 2004 Monsanto had approached China with a plan to ensnare ordinary employees, corporate executives, technical experts, and government officials. Those supported by international organizations and seed MNCs to pursue studies in the United States in the 1980s and 1990s had returned and become elite academicians, CPPCC members, or government officials carrying significant weight in the making of China's S&T policy and even controlling the media's orientation. Meanwhile, through funding research, conducting collaborative research, and having their agents appointed senior advisers at research institutes and universities, according to Tong, MNCs such as Monsanto had accelerated their localization and successfully embedded themselves in China's scientific, economic, and even political decision-making process.[44] Yet Tong seemed to ignore the fact Monsanto mainly engages in conventional breeding in China, having almost nothing to do with GM seeds, obviously because of restrictions in investment and variety registration that the Chinese government has imposed. Consequently, China only accounted for 1 percent of Monsanto's total revenue, which was $13.5 billion, in 2012.[45]

Similarly, simply because they were trained in the United States and received research grants from the Rockefeller Foundation and other American organizations, Chinese transgenic scientists, from Chen Zhangliang, Zhang Qifa, Li Jiayang, and Wang Dafang to Zhu Zhen, have been accused of being "representatives" (*dailiren*) of the American government and selling China's national interests as "traitors" (*hanjian*). Even students supported by Monsanto are future "soldiers" of the "collaborationist Chinese army" (*weijun*), referring to the Chinese fighting alongside the Japanese invader in the Sino-Japanese War between 1937 and 1945, with Zhang Qifa as their "commander."

The GMO conspiracy theorists also branded the China National Rice Institute, which collaborated with the International Rice Research Institute, a Rockefeller Foundation beneficiary in the Philipines, a *hanjian* organization.[46] The China International HarvestPlus project, part of the International HarvestPlus project aimed at biofortification, has been supported by the International Institute for Tropical Agriculture and the International Food Policy Research Institute, two members of the former Consultative Group on International Agricultural Research (CGIAR) consortium, which in turn is connected with the Rockefeller and Ford foundations. Therefore, Fan Yunliu, its one-time director, surely represented U.S. interests in her efforts to develop Bt rice and phytase maize.

The International Service for the Acquisition of Agri-biotech Applications (ISAAA) spares no effort to promote and advocate the commercialization of agricultural biotechnology and GM crops worldwide. As its main sponsor is none other than the Rockefeller Foundation, Clive James, its founder and current chairman of the board of directors, was called an American spy (*jiandie*) by Chinese GMO critics; so was Huang Jikun, a member of the ISAAA board of directors.[47] Once serving on DuPont Pioneer's advisory board and meeting Clive James in his capacity as vice minister of agriculture, Li Jiayang also conspired to cater to American interests and therefore should be dismissed.[48] Story even had it that Sun Zhengcai, then agriculture minister, flew Monsanto's corporate jet to visit his daughter at Cornell University while visiting the company in 2009.[49]

Zhou Li, an associate professor at Renmin University of China's School of Agriculture, went so far as to suggest that agricultural bio-technologists at key Chinese universities and research institutes "have willingly and effectively become members of the 'Unit 731,' " simply because they had received help from food and seed MNCs in bidding for research grants, publishing papers, conducting experiments, and engaging in commercial activities.[50] This accusation was outrageous because Unit 731 was the covert Japanese biological and chemical warfare unit that undertook lethal human experimentation in northeast China during the Sino-Japanese War; Zhou did not substantiate the allegations by actual facts as opposed to supposition. Coincidentally or not, the Golden Rice study was also referred to as a Unit 731 type of operation. And in late 2011, about two months before fleeing to the U.S. consulate in Chengdu seeking political asylum, Wang Lijun, the now infamous deputy mayor of Chongqing, attended a seminar on transgenic technology and human security, sponsored by the city's public security bureau. At the seminar, Wang associated the Americans' transgenic push in China with the agenda of Freemasons, the fraternal organization, which conspiracy theorists in the West have long accused of having the evil goal of world domination.[51]

Conspiracy theories tend to find guilt until innocence is proven, rather than the other way around, and one needs neither logical reasoning nor evidence to prove or refute such a theory. Psychologically, conspiracy theory is tempting and even flattering, because it easily satisfies its theorists' intellectual and moral positions. As all activities are doubtful, never simple, and connected, conspiracy theorists claim to be good at piecing together flickering fragments into a new picture that reveals a hidden, and mostly malicious, agenda. In the case of GM crops, according to the conspiracy theorists, Chinese transgenic scientists and MOA officials have been so keen to commercialize GM rice simply because of their willingness to help America to sabotage China. As suggested by Peng Guangqian, a major general of the People's Liberation Army (PLA) who also is a deputy secretary-general of China's National Security Forum, GMOs are taking shape as a Western conspiracy to supplant China's food security. "America has a long-term strategy to curb high-tech embargo on China, with the sole exception of transgenic technology. Interest groups such as Monsanto and DuPont have been unusually generous to dump GM crops to China. What is exactly behind this extremely abnormal activity on the American side? Is it a pie or a trap?" questioned Peng in *Global Times*.[52]

While organized skepticism is one of the normative requirements for science,[53] conspiracy theory is harmful in that it not only distorts facts and hurts innocent persons but also deconstructs the cornerstones on which a normal society is built. Conspiracy theorists do not judge persons or events on their own merits but assign a perpetrator's motive to them. Again, in the case of GM crops, the Chinese conspiracy theorists ignore the scientific foundations of transgenic technology by asking ambiguous questions such as "Why should people eat Bt rice that even insects do not eat," and spreading rumors to manipulate ordinary people's emotions. In their view, Syngenta AG, then world's third largest GM-seed producer, would soon be worthless son; Monsanto's bid for it was a conspiracy to fool ChemChina, whose offer of $43 billion to acquire Syngenta would be "a disaster for Chinese people and the entire human being" and "an ethnic genocide."[54] However, by effectively underscoring the populist posturing, conspiracy theorists always have an Achilles heel in addressing the public, who do not want to be treated as stupid, ignorant, uninformed, unintelligent idiots and have their own thoughts regarding these ambiguous theorists.

Finally, conspiracy theory against GM crops and foods has its market in China, thanks in large part to China's more open media and especially the proliferation of the internet and, most recently, social media outlets such as Weibo, a micro-blogging platform similar to Twitter, and WeChat, similar to Whatsapp and Facebook Messenger. Indeed, much of the misinformation and many unsubstantiated allegations have found their way not only to designated websites such as Utopia but also to internet portals such as Sina.com, as well as being published in newspapers and magazines. PLA Major General Peng Guangqian's comments, mentioned above, were published in *Global Times*, a populist and nationalist outlet of *People's Daily*, the Chinese Communist Party's mouthpiece. *Time Weekly*, a newsweekly published in Guangzhou, openly promoted the views of GMO conspiracy theorists such as Zhou Li, Renmin University's associate professor. Books using sensitive language to condemn GMOs and demonizing scientists and government officials have also been published.

Countermeasures from the Pro-GMO Camp

As the debates on transgenic technology and GMOs have been polarized and politicized, China's transgenic scientists have striven to meet the challenges from the anti-GMO camp. For example, to one of the questions that Greenpeace posed—whether the harmless conclusion from the three-month Bt-rice rat-feeding experiments, an internationally accepted practice, could mean the possibility of humans' safely consuming such rice for fifty years—Jia Shirong, the Bt-cotton inventor, answered by citing a line of Vladimir Lenin, the former Soviet leader: "One hundred wise men cannot answer one question raised by a stupid person."[55]

One weapon that the scientific community uses is to communicate to the public the knowledge about GM crops and foods by patiently explaining the science behind GMOs. The GMO MEP has allocated a significant amount of the funding for this purpose.[56] In 2011, MOA's Office of Agricultural Genetic Engineering Biosafety Administration and the Department of Science Popularization of the China Association for Science and Technology published a booklet on basic knowledge of agricultural GMOs. That same year also saw the launch of the Gene and Agriculture Net (www.agrogene.cn) website, sponsored by the China Society for Biotechnology, the China Society of Plant Physiology and Molecular Biology, the Crop Science Society of China, the China Society of Plant Protection, and China Agricultural Biotechnology Society. The website aims to serve as a platform through which scientists and the media raise the public's awareness of agricultural biotechnology. It has "Q&A," "Rumors and Truth," and other sections, addressing public concerns over transgenic technology and dissipating rumors and unfound attacks on the technology. Users are encouraged to ask questions on the website, which then invites scientists to respond.

The pro-GMO camp has also organized GM-food tasting sessions. On August 17, 2014, the day when the biosafety certificates issued to Huazhong Agricultural University's insect-resistant Bt rice expired, more than 600 participants tasted Bt rice in twenty-two Chinese cities, including Beijing, Shenzhen, Shanghai, Xiamen, Nanjing, Hangzhou, Wuhan, Chengdu, and Xian.

Thus far, some one hundred GM-food tasting sessions have been held in China. The idea of tasting GM foods came from Fang Zhouzi, the science writer. On May 19, 2011, Fang Xuanchang, a journalist who is now the editor-in-chief of the recently established Gene and Agriculture Net, organized the first tasting at a Guangzhou restaurant, where twenty-four persons tried Bt rice provided by Huazhong Agricultural University, as well as foods made with edible oils extracted from GM soybeans, and GM papaya, a local favorite approved for commercialization. Four days later, Fang Zhouzi organized the first Beijing Bt-rice tasting event, which attracted thirty participants. One of the largest Bt-rice tastings, held on July 13, 2011, was sponsored by *Beijing Science and Technology Daily*; attendees included Chen Junshi, an academician of the Chinese Academy of Engineering working at the Chinese CDC, and Lin Yongjun from Zhang Qifa's lab at Huazhong Agricultural University. On the same day, similar tasting sessions were organized in Kunming, Nanning, and three other Chinese cities.

According to Fang Xuanchang, the Bt-rice tasters have no specific agenda other than expressing their willingness to try Bt rice. They intend to show that Bt rice poses no threat to food safety and treat the tasting as an event for communicating and experiencing the science. But these volunteer Bt-rice tasters have been accused of being bribed by the GM interest group, thus giving them vested interests as well. The reaction from the anti-GMO camp was especially manifested by a comment that reporting the Bt-rice tasting activity is illegal, as is the spreading of GM material—Bt rice—by Huazhong Agricultural University.[57]

The anti-GMO camp was frantic. Before the August 17, 2014, nation-wide Bt-rice tasting, a private event, Chongqing police confiscated Bt rice and T-shirts prepared for the occasion and demanded that participants not attend the tasting session in Chengdu, an adjacent city. Two days before the tasting, Sohu.com, one of China's leading internet portals, deleted Fang Zhouzi's four blog posts and banned him from posting on the site for forty-eight hours for hyping GMOs, criticizing the Chongqing police, and reminding fellow Bt-rice tasters of the attendance at the event. In his remarks at the Beijing Bt-rice tasting event, which attracted some one hundred persons, Fang emphasized that the date was chosen to show "the power of science"; the persons at the event were not eating their "last meal," but looked forward to the time "when GM foods become part of our daily lives."[58]

The battle between the pro- and anti-GMO camps continues.

Summary and Discussion

This chapter details a delicate and troubled relationship between the transgenic research community and the public in China in dealing with issues pertaining to GM crops and foods in general and Bt rice in particular. Although their knowledge base is expanding, the public as a whole still may not quite understand the science carried out at the frontier of agricultural biotechnology. They are likely being influenced by communications from scientists who are more knowledgeable about the subject. However, Chinese agricultural biotechnologists have not done a good job dialoguing with ordinary citizens,[59] thus giving rise to a conflicting and negative sentiment about GMOs. While concerns over the safety of GM foods for human consumption and over the potential impact of GMOs on the environment and biodiversity, such as cross-pollination and gene flow, may have polarized GM scientists and the public, it is the global and domestic anti-GMO populism that tends to politicize and ideologize the debates.

At a time when domestic NGOs—other than Utopia, which had its political agenda—generally did not take a stand on the issue, Greenpeace stepped in to fill the gap, not only provocatively raising awareness of biosafety and food safety associated with GMOs among Chinese citizens and influencing policy, but also directing a fusillade of criticism at transgenic scientists. Allowing Greenpeace to enter China and orchestrate a full-fledged and extensive campaign against GM crops and foods and allowing Chinese media to publish its anti-GMO manifesto and stories, to be discussed in the following chapter, indicate that Chinese society has become more open and pluralistic. But it may also signal ambiguity on the part of the leadership toward GMOs or political sympathy with the activism within the hierarchy of the party-state. Indeed, NGOs in China have to walk a fine line in their campaigns on particular causes as the communist party generally holds a negative attitude toward them, particularly those with foreign connections, as is reflected in its most recently enacted law heightening layers of control and oversight.[60] But the party has also become open-minded toward environmental-protection NGOs; for example, it did not intervene when domestic environmental NGOs led protests against using hydroelectricity and nuclear energy to increase the energy supply.[61] One wonders, therefore, whether tolerance toward Greenpeace's activism had been sanctioned by the government agencies in charge of environmental protection and biosafety in which they were unable to play a direct role. When disputes arose around imports of GM soybeans and corn to China, the voice from an international NGO might serve as the excuse or the best defense for the government, fending off accusations by the United States and other countries regarding China's invocation of biosafety to block such imports. If this was the case, China might be hypocritical in following an international trend and accommodating domestic pressure for the "right to know." Calling for the labeling of GM foods so that consumers can make an informed choice may also be part of a strategy favored by a different government agency. If so, once they are ready to commercialize their own GM varieties, Chinese scientists are likely to encounter not only Greenpeace's activism but also government's hesitation. Moreover, as the media, including the internet, are controlled, monitored, scrutinized, and censored in China, it is also remarkable that newspapers and posts on the internet could have been tolerated in advancing a conspiracy theory against government policy, spreading misinformation, and demonizing scientists and government officials. Thus, it is understandable why the government has been promotional on GMOs, at least on the research side,[62] while seemingly precautious by not censoring views significantly different from its policy.

The anti-GMO campaign may also represent public antagonism toward scientists and the establishment. Indeed, in China's reform and open-door era, scientists have become an advantaged group, not only succeeding professionally but also benefiting from various business and political opportunities. They become part of the "the power elite," using the term coined by sociologist C. Wright Mills some sixty years ago to describe a relatively small, loosely connected group of individuals who dominated America, taking all the wins in science, business, and even politics.[63] This also explains why transgenic scientists have received complaints from those engaged in other lines of research for being marginalized.

However, the activism began to affect actual policy. "As is true for many public policies, changes in risk regulations typically have their origin in some kind of triggering mechanism," such as "a major accident, catastrophe, or highly visible policy failure, new reports or studies, an influential book, stories in the media, and/or a public campaign waged by activists."[64] Indeed, Chinese consumers first heard of transgenes (*zhuanjiyin*) and GM foods from Greenpeace's campaign for the consumer's "right to know" in its battle over Nestlé in 2002, which is emblematic of the broader fight on GMOs. Consumers' already ambiguous views toward GM foods only became more confused when the war of words and actions between the pro- and anti-GMO camps around Bt rice ratcheted up after 2009. While it is difficult to conclude that strong and adamant activism has succeeded in stalling Chinese government's moves toward commercializing agricultural biotechnology, it also is difficult to deny that the developments discussed in this chapter seem to indicate a turning point when China started to see its policy of GM crops and foods gradually shift from promotional, or at least permissive, to precautionary options.

CHAPTER VI

The Chinese Media and Changing Policy

I n today's increasingly technologically sophisticated society, the internet and social media have become critical channels of sharing and disseminating information. However, traditional media such as newspapers and television, while losing audience and revenue, are still resilient and remain purveyors of information beyond personal experience and education. Newspapers, especially the prominent ones, are read by political and business elite and the well-educated, who in turn use their influence to help set policy agendas, prompt debates on specific issues, and affect public opinions and policymaking. Given their significance for public opinion and policy outcomes, newspapers are both producers of discourse and "a site on which various social groups, institutions, and ideologies struggle over the definition and construction of social reality."[1] Meanwhile, like all media, newspapers present news and construct a reality through the filter of journalistic language and imagery. Thus, it is inevitable that journalists, in narrating facts and information, highlight certain viewpoints while marginalizing or ignoring others,[2] organizing or framing events into a coherent story, and depicting one perspective as more favorable or reasonable than others.[3]

Science journalists are no exception.[4] In particular, discussion of transgenic technology has not been confined to the scientific community and policy community but increasingly has been extended to and penetrated the media. By framing the risks and/or benefits of GM crops and foods, the media not only present opinions of various stakeholders but also help shape policy formulation and change. For example, an analysis of the relationship between the British press coverage of biotechnology between 1973 and 1999 and public perception between 1996 and 1999 concluded that the debates over GM crops and foods had fostered the contrast between "desirable" medical (red) and "undesirable" agro-food (green) biotechnology in the newspaper-reading public, thereby not only shielding red biotechnology from public controversy but also ushering in a realignment of the British regulatory framework in 2000.[5] Similarly, a study of the coverage of biotechnology in the *Times* and the *Sunday Times* in the United Kingdom and the *Washington Post* in the United States between 1990 and 2001 found that medical biotechnology was framed more positively and agricultural biotechnology more negatively, a finding that had held true over the study period. Such coverage helped shape public perception of the two biotechnologies: positive toward medical applications but more negative (or ambivalent) toward agricultural ones.[6]

In Beijing, China's capital, according to an early twenty-first-century study, the main channels through which consumers acquired information about GM soybeans were television (45.5 percent) and newspapers and magazines (35.3 percent); a similar online survey indicated that newspapers and magazines (35.2 percent) and broadcasting and television (27.3 percent) were key sources for information, followed by the internet (23.6 percent).[7] The coverage of GM foods in two major Chinese newspapers—*People's Daily* and *Guangming Daily*—between January 2002 and August 2011 was mostly supportive of transgenic technology R&D programs and the adoption of Bt cotton, although the two newspapers might not fully reflect public opinion.[8]

This chapter takes a more comprehensive and longitudinal approach to explore how key issues pertaining to GM crops and foods have been reported in some of the leading Chinese newspapers. In particular, it will start with a systematic examination of the coverage in three dailies and two weeklies, followed by case studies looking more specifically at how these newspapers covered key GMO-related events. The chapter will speculate on how likely the changing coverage over time affected public perception of and thus policy toward GMOs.

Data Collection

After consulting several Chinese science journalists with experience covering transgenic technology and GMOs regarding the selection of representative print media outlets for this study, I decided to focus on the following five:

- *People's Daily* (Renmin Ribao), the official newspaper of the Chinese Communist Party (CCP) with a circulation of 2.8 million in 2013, which authoritatively conveys party's policies.
- *Science and Technology Daily* (Keji Ribao), published by the Ministry of Science and Technology (MOST), the Chinese government bureaucracy in charge of the scientific enterprise. As MOST formulates S&T development plans and policies and organizes and implements national S&T programs, including those related to agricultural biotechnology, this widely circulated newspaper—250,000 copies in 2013—represents the views of MOST as well as those of the research community.
- *Farmer's Daily* (Nongmin Ribao), the newspaper of and for Chinese farmers. With a circulation of 700,000 in 2013, the newspaper is a joint publication of the State Council's Research Center for Rural Development and the Research Center for Rural Policy under the CCP's Central Committee, therefore presumably conveying an official voice and that of Chinese farmers.
- *Southern Weekend* (Nanfang Zhoumo), one of the publications of the Southern Media Group, a Guangdong provincial government-owned media group. With a circulation of 1.28 million in 2013 and applauded by the *New York Times* as "China's most influential liberal newspaper,"[9] *Southern Weekend* is well known for its investigative journalism, often testing the limits of free speech in the country.
- *Outlook Weekly* (Liaowang Zhoukan), published by Xinhua News Agency, China's state news agency. It is another official media outlet, with a circulation of 470,000 in 2013, aiming to provide in-depth analysis of topics that affect Chinese citizens and public policy.

Articles published in the first four—*People's Daily*, *Science and Technology Daily*, *Farmer's Daily*, and *Southern Weekend*—are retrievable from China National Knowledge Infrastructure (CNKI). An e-publishing platform established in 1996, CNKI contains journals, newspapers, dissertations, proceedings, yearbooks, and reference works, among others, published in China. Among its four databases is China Core Newspaper Database (CCND), which has collected core Chinese newspapers since 2000. Using keywords, we retrieved all articles published in the four newspapers on "transgene" (*zhuanjiyin*) and "GM foods" (*zhuanjiyin shiping*) between January 1, 2000 and December 31, 2012. We then used a similar search strategy, via a VIP resource integration service platform, to retrieve all "transgene" and "GM foods" related articles published in the *Outlook Weekly* during the same period. A total of 420 articles were collected. Of these, 220 articles were from *Science and Technology Daily*, accounting for more than half of the total, as transgenic technology and GM crops and foods are mainly an S&T-related subject; *Farmer's Daily* published 115 stories as the subject is closely related to farmers' interests, 55 were from *People's Daily*, and 15 articles each were from *Southern Weekend* and *Outlook Weekly*. Of the articles, 373 were news stories—203, 100, 48, 12, and 10 in *Science and Technology Daily*, *Farmer's Daily*, *People's Daily*, *Southern Weekend*, and *Outlook Weekly*, respectively—and the rest were commentaries.

Characteristics of Coverage

Trend

Measured by number of stories, different newspapers show a changing but not random coverage of transgene and GM foods over time (table 6.1 and figure 6.1). All newspapers adhered to a similar trend, with two upsurges over the studied period—one between 2002 and 2004 and the other between 2009 and 2010. Cyclical coverage seems to be event-driven, trending up in response to a major event and down sharply until the appearance of another important event.

The year 2002 was a turning point for GMOs in China. When ordinary Chinese still did not know much about what GMOs would bring to their life, GM soybeans started to arrive from overseas. As discussed in chapter 4, at that time, the State Council and the Ministry of Agriculture came up with policy documents, aiming to establish a biosafety regulatory regime suitable for the development of agricultural GMOs. Therefore, the coverage focused not only on the progress at the forefront of research of GMOs but also on biosafety and policy measures in various countries. In 2003, as noted in chapter 5, Greenpeace introduced "transgene," the vocabulary, to Chinese by way of its campaign against Nestlé, alleging that the company did not properly label products containing GM ingredients sold in China. But the media coverage did not pick up the case until late 2003 and early 2004.

TABLE 6.1

Number of Reports on GMOs (2000–2012)

Year	People's Daily	Science and Technology Daily	Farmer's Daily	Southern Weekend	Outlook Weekly	Total
2000	3	9	5	0	1	18
2001	1	13	12	0	0	26
2002	3	33	23	0	2	61
2003	9	15	13	0	1	38
2004	3	17	3	3	3	29
2005	6	21	2	1	1	31
2006	2	17	4	0	1	24
2007	3	22	2	0	0	27
2008	6	13	8	0	0	27
2009	3	14	18	1	1	37
2010	6	15	14	2	5	42
2011	5	11	6	5	0	27
2012	5	20	5	3	0	33
Total	55	220	115	15	15	420

Sources: Analysis based on articles reporting "transgene" (zhuanjiyin) and "GM foods" (zhuanjiyin shiping) in five newspapers retrieved and downloaded from the China Core Newspaper Database (CCND) of China National Knowledge Infrastructure (CNKI) and a VIP integration service platform.

Afterwards, Chinese began to think carefully about what GMOs meant to them in terms of benefits and risks and whether they could bear the consequences associated with such risks, if any, corresponding to a transition of Chinese policy from promotion in research to permission and even precaution in trade and commercialization. The coverage then proceeded at a low ebb for a number of years. Science journalists seemed frustrated as to how to report GMOs, which explains why the inclusion of developing new varieties of GMOs in the Mega-Engineering Programs of the Medium- and Long-Term Plan for the Development of Science and Technology in 2006 did not lead to immediate media frenzy or indifference. Some newspapers only carried interviews of or commentaries by scientists who used the policy document, MLP, as evidence to emphasize government's willingness to push forward transgenic technology.

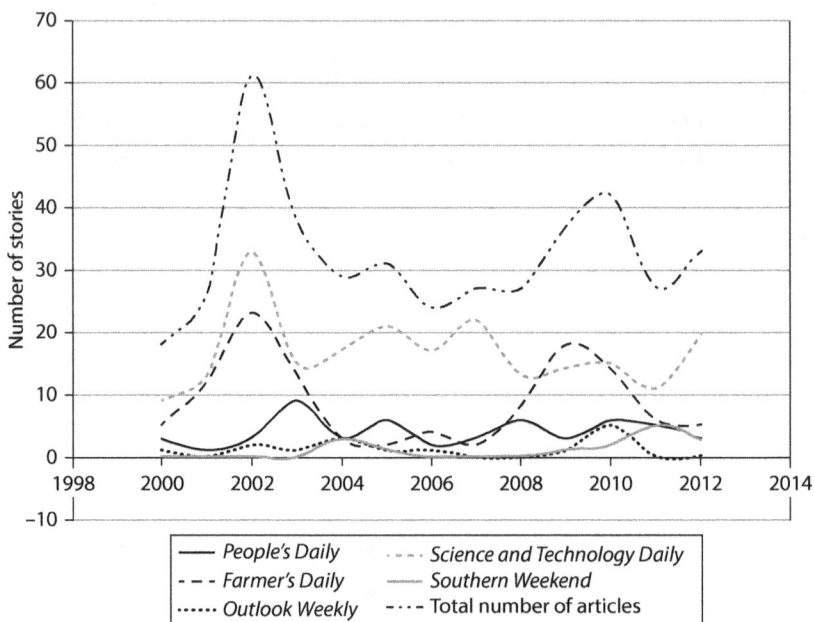

Figure 6.1 The changing numbers of GMO stories in select Chinese newspapers (2000–2012).

Sources: Analysis based on articles reporting "transgene" (*zhuanjiyin*) and "GM foods" (*zhuanjiyin shiping*) in five newspapers retrieved and downloaded from the China Core Newspaper Database (CCND) of China National Knowledge Infrastructure (CNKI) and a VIP integration service platform.

In August 2009, after years of inaction, MOA finally issued biosafety certificates to two strains of Bt rice and one strain of phytase maize. The news was made public in early December and started to receive coverage thereafter. Although members of the biotechnology research community hailed the certification as a milestone in their decade-long efforts, the prospect of rice, a staple food, being genetically altered concerned an increasing number of ordinary Chinese about its possible negative effects. Coincident or not, the coverage reached a second peak. In particular, *Farmer's Daily* quite actively and positively reported the story, but *Science and Technology Daily* was unexpectedly quiet. Toward the end of 2012, the reporting in *Science and Technology Daily* started to pick up again, conveying promotional views of the scientific community amid an overall precautious environment toward commercializing the Bt rice.

A piece of S&T policy and its relevance to the public's life may be an important factor determining media attention. The first peak in coverage by the newspapers under study corresponded to the case of Nestlé's failing to label its GM-containing products. The second peak did not arrive when MLP was released; despite its devotion to GMOs, the GMO MEP is remote from the life of ordinary citizens. It was the news of the issuing of biosafety certificates to Bt rice and maize and the higher likelihood that Bt rice would soon appear on Chinese dining tables, thus affecting everyday Chinese life, that brought increased coverage in the newspapers. Meanwhile, as the party's mouthpiece, Chinese newspapers have to closely follow party policy. Generally, newspapers tend to delay reporting an issue when there is conflict over policy, waiting for the dust to settle before starting to cover it, even though doing so undermines journalism's principle of timeliness.

Placement

The position of a news story often reflects its relative significance. Normally, a story on the front page is more important, likely to be read by more readers, and therefore more influential. In general, GMO stories seem not important in the newspapers under study. While not making onto the front pages of *People's Daily* and two weeklies, GMO stories did appear on page one of *Science and Technology Daily* (twenty-one stories) and *Farmer's Daily* (seven stories), two newspapers of more immediate relevance to the subject covered.

Besides front pages, GMO stories were mostly published on pages or in columns devoted to international and science news, showing these newspapers' interest in the new technology and its worldwide development trends (table 6.2). For example, *Science and Technology Daily* focused on the international development of transgenic technology as the development of GMOs in other countries might help shape China's policy. GMO stories were also likely to appear on the pages of agricultural and economic news. *Farmer's Daily* carried more stories, mainly about Bt cotton, on its agriculture page. Indeed, Bt cotton is the variety closest to Chinese farmers who, as beneficiaries of the transgenic technology, look forward to growing other GM crops. This confirms findings from a study of GM crop and food coverage in *People's Daily*, *Science and Technology Daily*, *China Youth Daily*, and *Jinghua Times* between 2002 and 2006. According to that study, *Science and Technology Daily* tended to pay more attention to the development of GMOs, with front-page stories accounting for some 10 percent of the total; *People's Daily* was more

TABLE 6.2

Placement of the Coverage of GMOs (2000–2012)

	People's Daily	Science and Technology Daily	Farmer's Daily	Southern Weekend	Outlook Weekly	Total
Front page	0	21	7	0	0	28
International news	12	59	11	0	0	82
National news	12	16	8	0	0	36
Political news	2	5	3	5	10	25
Economic news	8	4	3	1	0	16
Science news	13	38	27	7	5	90
Agriculture news	0	13	45	0	0	58
Others	8	64	11	2	0	85
Total	55	220	115	15	15	420

Sources: Analysis based on articles reporting "transgene" (zhuanjiyin) and "GM foods" (zhuanjiyin shiping) in five newspapers retrieved and downloaded from the China Core Newspaper Database (CCND) of China National Knowledge Infrastructure (CNKI) and a VIP integration service platform.

likely to place policy-oriented and comprehensive reports on Bt cotton on domestic pages, news about commercialization and application of Bt cotton on economic pages, and explanatory and scientific stories on S&T pages.[10]

Main Themes

As noted, GMOs can be framed in terms of benefits and/or risks. A study shows that coverage in U.S. and UK newspapers tended to frame GM crops in terms of environmental risks rather than benefits.[11] Studies of Chinese newspapers find a prevalence of themes about benefits over risks in both *People's Daily* and *Guangming Daily* between January 1, 2002 and August 31, 2011;[12] specifically, between 1995 and 2006, about three-quarters (73.5 percent) of the GMO coverage in *People's Daily* was framed in the progress/economic perspective, and 59 percent dealt with various other interests.[13] The three most frequently encountered thematic categories in the newspapers under study include progress and achievements, food safety, and debate

TABLE 6.3
Major Themes of the Coverage (2000–2012)

	People's Daily	Science and Technology Daily	Farmer's Daily	Southern Weekend	Outlook Weekly	Total
Progress and achievement	12	72	36	1	3	124
Food safety	12	51	16	2	2	83
Policy and governance	10	37	16	7	6	76
Debate on risks	7	48	9	7	8	79
Benefits	8	20	18	0	0	46
Commercialization	0	8	18	1	1	28
Market system	8	17	9	1	1	36
Comparison	0	2	1	0	0	3
Plant	2	24	20	2	0	48
Trade war	4	5	5	0	0	14
History	0	6	2	0	0	8
Staple food modified	0	0	1	1	0	2
Total	63	290	151	22	21	547

Sources: Analysis based on articles reporting "transgene" (zhuanjiyin) and "GM foods" (zhuanjiyin shiping) in five newspapers retrieved and downloaded from the China Core Newspaper Database (CCND) of China National Knowledge Infrastructure (CNKI) and a VIP integration service platform.

on risks, which collectively comprise almost half (49 percent) of the total (table 6.3). In particular, *People's Daily* was concerned more about food safety, *Science and Technology Daily* and *Farmer's Daily* attended to progress and achievements at the frontier of research, and *Southern Weekend* and *Outlook Weekly* analyzed the overall risks of GMOs.

On average, a story has 1.3 themes, with the three dailies having relatively focused themes and the two weeklies covering a wide variety of themes. As the party's newspaper, *People's Daily* covered the smallest number of themes and was unequivocally supportive of GMOs; it almost ignored the issue of risks, as measured by themes as well as quality, content, and position of the stories and commentaries.[14]

The themes are closely connected and cannot be easily separated. Progress and achievements on GMO research were more likely to be covered along with commercialization; the development of a regulatory regime and research and administration on safety in various countries were reported simultaneously; reports of consumers' "right to know" appeared together with the themes of safety and regulations. However, the commercial growing of various GM crops was not reported together, reflecting a reality that China has only approved the commercialization of Bt cotton and papaya and other GM crops have yet to reach the stage of large-scale production.

While there had been no obvious thematic changes over time in different newspapers, the coverage in *People's Daily* in 2005 was centered on food safety and consumers' "right to know," and the newspaper had showed a rising interest in regulations in different countries after 2010. Reporting of progress and achievements at the frontier of research was most significant in *Science and Technology Daily* in 2007 and *Farmer's Daily* in 2009. Overall, reports of food safety and the debates on risks came up mostly around 2002; the year 2009 saw the most coverage of progress and achievements in research, after which the coverage gave increasing attention to how different countries regulate and administrate the risks of GM crops and foods.

Tone of Coverage

Now, I move on to an analysis of how Chinese newspapers approached their GMO stories by looking at their sources and headlines. I am interested in whether sources and headlines focused on the benefit/positive or risk/negative frame of the GMO debates, with neutral being the sources or headlines framed in a balanced way. I also will examine the overall tone of the stories.

Source Utilization

In order to provide readers with comprehensive information about an issue for their judgment, a newspaper should strive to maintain the ideal of neutrality and unbiased reporting by getting access to multiple sources, balancing diverse viewpoints, presenting all sides fairly, and distinguishing news reporting from editorial opinion. A study of the coverage of avian flu in Australian print, radio, and television media, for example, finds a strong relationship between the quality of coverage in a news medium and the number of sources consulted.[15] Reporting two sides of an issue can enhance objectivity.[16] This is especially the case for covering a controversy, where quoting sources representing opposite sides is more important for maintaining neutrality and integrity,[17] and where the public feel their input through the media, in a broad sense, being appreciated in policymaking.[18] Bearing these factors in mind, I look at how sources had been utilized in the coverage of GMOs in the five Chinese newspapers. First, the number of sources cited in a news story is used as a proxy for the effectiveness of source utilization. There were a total of 757 sources cited in these newspapers, or each news story sought for comments, on average, from slightly more than two sources (table 6.4).

Second, the views expressed by the sources were overwhelmingly supportive of the development of GMOs in China. Close to two-thirds (63.28 percent) of the sources made their comments in a positive/supportive/promotional tone, one-sixth (17.44 percent) appeared to be negative/antagonistic/precautious about the development of transgenic technology, and the rest took a neutral position. This echoes findings of biotechnology reporting between 1970 and 1999 in the United States, where there was an obviously supportive orientation toward the technology.[19]

TABLE 6.4

Sources Utilized in News Stories (2000–2012)

	Negative	Neutral	Positive	Total	Percent
Industries	20	24	54	98	12.95
Officials	25	35	83	143	18.89
Scientists	47	61	321	429	56.67
Humanists	14	18	17	49	6.47
Representatives of consumer organizations	26	8	4	38	5.02
Total	132	146	479	757	100.00
Percent	17.44	19.29	63.28		

Sources: Analysis based on articles reporting "transgene" (zhuanjiyin) and "GM foods" (zhuanjiyin shiping) in five newspapers retrieved and downloaded from the China Core Newspaper Database (CCND) of China National Knowledge Infrastructure (CNKI) and a VIP integration service platform.

Third, who the sources are is as important as how the sources give their views. The main reason for the seeming skewness toward positive sources in the newspapers under study may be the overrepresentation of scientists (56.67 percent) as sources that the journalists sought for comments. It is understandable that scientists involved in the research and commercialization of GM crops tended to be more positive (74.8 percent) on the issue than other stakeholders. Corporate representatives were also positive, as were government officials; more than half of both groups expressed views in support of the development of transgenic technology in China. Humanities scholars were more evenly balanced among positive, neutral, and negative views. Consumer organizations, the smallest group among the sources, held negative views on GMOs by a large margin (68.4 percent).

Indeed, the existence of an "inner circle" for science journalists—scientists—explains why science stories are similar.[20] Despite differences among the media, science journalism tends to focus on the same issues, cite the same sources, and interpret information in similar terms.[21] According to the so-called information- or knowledge-deficient model of science communication, the public's uncertain and skeptical views toward emerging technologies such as biotechnology and nanotechnology are generated primarily by a lack of sufficient knowledge about science and the relevant subjects. By providing adequate information, scientists can help narrow the knowledge gap that exists in the public.[22] As information providers, scientists tend to view the media as a conduit or pipeline for communicating scientific findings to the public for them to consume, digest, understand, and, where appropriate, take action.[23] In some extreme cases, scientists even assume that the media should educate the public about science and risks.[24] Chinese science journalists also often look to scientists for comments on the debates surrounding GM crops and foods.[25] In a series published in *People's Daily* on November 27, 2008, for example, three of the six stories involved interviews with transgenic scientists.[26] Again, this is similar to the situation in the United States, where

"the pro-biotechnology tone of media coverage and the dominance of science, industry, or government sources are also attributable to journalistic preferences, as the media have relied heavily on these sources for technical information and routine channel news. The fact that the source-journalist interaction has benefited pro-biotechnology sources at the expense of dissenting interests is promoted by a favorable run of history in the United States."[27]

Headlines

Apart from issue/theme frames, a story's headline, though condensed, usually contains the main idea of the story and therefore is a good "window" showing a newspaper's stance on the covered topic. For the sake of analysis, headlines of the GMO stories in the five newspapers were categorized as positive, negative, or neutral. A positive or negative headline depended upon whether it approached GMOs in a benefit or risk frame, with a neutral headline framing neither aspect or considering both frames. The analysis was limited to the news stories.

Contrary to the conventional wisdom that a technology with uncertainty is likely to be reported with neutral headlines, the five Chinese newspapers under study tended to report GMOs in a very supportive tone, with 65.7 percent of the headlines being positive (table 6.5). This is mainly attributed to the tone of headlines in the three dailies—*People's Daily*, *Science and Technology Daily*, and *Farmer's Daily*; the headlines in the two weeklies were mostly neutral.

Overall Tone

As a whole, over the period, the tone of reporting was quite balanced (table 6.6). Most news stories reported GMOs neutrally as an early negative tone gave way to a positive tone in 2002 and 2003, then turned to slightly negative in 2010 and 2011. Only *People's Daily* showed a clear and increasingly supportive tone in reporting on GMOs. (The coverage in *Southern Weekend* and *Outlook Weekly* did not show a pattern as each published a limited number of stories over the study period.) The period between 2002 and 2003 was special in that stories with a positive tone reached their highest, significantly more than those with the negative and neutral tones. However, the release of MLP in 2006, which included a GMO MEP, saw neither an increasing number of news stories, as noted, nor the more positive tone expected. A study of the coverage of GMOs in authoritative *People's Daily* and *Guangming Daily* between 2002 and 2009 indicates that reports were largely supportive of the transgenic technology R&D programs and Bt cotton and none expressed negative views, although slightly more than half of all stories (51.9 percent) were neutral.[28] Indeed, the political leadership

TABLE 6.5

Headlines of News Stories (2000–2012)

A. By Newspaper

	People's Daily	Science and Technology Daily	Farmer's Daily	Southern Weekend	Outlook Weekly	Total
Negative	3	21	5	5	1	35
Neutral	15	55	15	4	4	93
Positive	30	127	80	3	5	245
Total	48	203	100	12	10	373

B. By Year

Year	Negative	Neutral	Positive	Total
2000	1	6	9	16
2001	1	6	15	22
2002	9	13	29	51
2003	1	16	19	36
2004	4	11	12	27
2005	3	6	20	29
2006	3	5	16	24
2007	1	4	20	25
2008	1	6	17	24
2009	2	4	30	36
2010	2	3	27	32
2011	2	3	14	19
2012	5	10	17	32
Total	35	93	245	373

Sources: Analysis based on articles reporting "transgene" (zhuanjiyin) and "GM foods" (zhuanjiyin shiping) in five newspapers retrieved and downloaded from the China Core Newspaper Database (CCND) of China National Knowledge Infrastructure (CNKI) and a VIP integration service platform.

TABLE 6.6

Tone of News Stories (2000–2012)

Year	Negative	Neutral	Positive	Total
2000	2	3	11	16
2001	4	1	17	22
2002	7	5	39	51
2003	3	8	25	36
2004	4	8	15	27
2005	3	2	24	29
2006	3	3	18	24
2007	1	0	24	25
2008	2	3	19	24
2009	1	2	33	36
2010	3	4	25	32
2011	3	1	15	19
2012	3	5	24	32
Total	39	45	289	373

Sources: Analysis based on articles reporting "transgene" (zhuanjiyin) and "GM foods" (zhuanjiyin shiping) in five newspapers retrieved and downloaded from the China Core Newspaper Database (CCND) of China National Knowledge Infrastructure (CNKI) and a VIP integration service platform.

has been positive toward the development of transgenic technology, as seen in the statements of Central Documents No. 1, discussed in chapter 3, although there was a subtle change in wording over time. Therefore, official media were less likely to deviate from the official policy.

Opinion Pieces

A newspaper's influence is felt most strongly through its opinion pieces. In particular, opinions in elite newspapers can significantly affect the public.[29] Only thirty-two commentaries in the newspapers under study have known information on their authors (table 6.7), of whom twenty-eight were scientists. These commentaries were overwhelmingly supportive of the development of GMOs in China, with the exception of two: one was about the development of the domestic soybean industry and the other

TABLE 6.7

Tone of Opinion Pieces (2000–2012)

Year	Headline			Overall			Total
	Negative	Neutral	Positive	Negative	Neutral	Positive	
2000	0	2	0	0	1	1	2
2001	0	1	3	0	0	4	4
2002	0	5	5	3	0	7	10
2003	1	0	1	0	1	1	2
2004	1	1	0	1	0	1	2
2005	1	0	1	0	0	2	2
2006	0	0	0	0	0	0	0
2007	0	1	1	0	0	2	2
2008	1	1	1	1	0	2	3
2009	1	0	0	1	0	0	1
2010	0	5	5	4	1	5	10
2011	0	4	4	0	0	8	8
2012	0	0	1	0	0	1	1
Total	5	20	22	10	3	34	47

Sources: Analysis based on articles reporting "transgene" (zhuanjiyin) and "GM foods" (zhuanjiyin shiping) in five newspapers retrieved and downloaded from the China Core Newspaper Database (CCND) of China National Knowledge Infrastructure (CNKI) and a VIP integration service platform.

about possible health effects of GM foods. Huang Dafang of the Institute of Biotechnology under the Chinese Academy of Agricultural Sciences (CAAS) penned four commentaries in *People's Daily*, *Science and Technology Daily*, and *Farmer's Daily*. Huang's colleague Jia Shirong, Peng Yufa of the CAAS Institute of Plant Protection, Zhang Qifa of Huazhong Agricultural University, and Yan Yang of South Dakota State University in the United States each authored two opinion pieces (Yan's supportive pieces in *Southern Weekend* were about the development of transgenic technology in the United States). Three commentaries with authors from the agricultural industry were also supportive, as were two by government officials. Only one commentary in *Outlook Weekly*, authored by a social scientist, contested the way in which policy on GMOs has been made.

Coverage of Important Events

As well as the tone of the coverage, the selection of events to report and the way of portraying/framing the events might have a profound impact on the public's perceptions of these events, and even influence policymaking.[30] Here I examine how two key events were covered in the five Chinese newspapers.

Failure to Label the Nesquik

Since entering the twenty-first century, there has been a dramatic increase in the media coverage of GM crops and foods in China, corresponding to their introduction.[31] Whereas the high-profile case in which the Swiss company Nestlé did not label its GMO-containing Nesquik sold in China triggered the first GMO controversy, as discussed in chapter 5, it was Greenpeace, the international NGO, that brought ordinary Chinese people's attention to the concept of transgene. With an effective media strategy that has been integrated into its campaign against GMOs in China, its staff members willingly communicate with the media and mobilize the media to serve its campaign, thus affecting public opinion toward GMOs and making the issue politically salient.[32] Indeed, Zhu Yanling's case against Nestlé became a media fiasco, receiving blanket coverage from more than thirty media outlets throughout the nation, including Chinese state broadcaster China Central Television (CCTV), in late 2003 and early 2004:

- "Nestlé Adopts a Double Standard on GM in Europe and China," *China Youth Daily* (Beijing), December 16, 2003
- "A Chinese Consumer Challenges a Fortune 500 Company: Opening Nestlé's Silent Door," *Youth News* (Shanghai), December 16, 2003
- "Zhu Yanling: I Sued Nestlé," *Southern Weekend* (Guangzhou), December 25, 2003
- "Nestlé Violates Moral Law," *Business Watch Magazine* (Xiamen), No. 1, 2004
- "Two Cases and a Perseverant Customer Who Sued Nestlé for Double Standards," *People's Daily* (Beijing), January 5, 2004

As the nation was caught up in an unusual frenzy over the safety of GM foods, some newspapers seemed to act as "amplification stations,"[33] magnifying risk-related issues so as to cast a precautious impression on GM foods. According to a study of the coverage of the case in four Chinese newspapers, *People's Daily* and *Science and Technology Daily* did not report the case (my study shows that *People's Daily* did publish a story). However, the journalists at *China Youth Daily* and *Jinghua Times* mainly used Zhu and her attorneys as sources without consulting biotechnologists and food-safety experts, and therefore their stories lacked important technology and food-safety angles. Neither newspaper interviewed social scientists or humanists, who might have a different perspective on consumers' "right to know" and to choose, nor did they interview relevant government officials to verify one of Nestlé's key arguments that it did not need to label its products according to Chinese regulations on GM foods.[34]

The coverage made an increasing number of Chinese consumers aware, for the first time, that they might have consumed GM foods. Some consumers went so far as to vent their anger on China's leading internet portals. However, along with the public's heightened awareness of the existence of GM foods in China was a lack of knowledge beyond that level.[35] A survey in Nanjing around that time indicates that 62.6 percent of the respondents knew "a little" and 32.1 percent knew "not at all" about GM products; but those without knowledge of whether GM products would have adverse effects on human health (47.3 percent) or the environment (50.4 percent) nonetheless believed that GM corn posed a potential risk to human health (46.6 percent) and overwhelmingly agreed that GM foods should be labeled (81.7 percent).[36]

Balanced information may well change the public's perception of and attitude toward GM foods. According to a three-stage survey of 480 respondents, also conducted in Nanjing around the time, 272 (56.7 percent) knew nothing about GM foods. Even among the 208 who had heard of GM foods, about 74 percent had only "heard the term" but had "no idea" of the benefits and risks associated with them; 99 respondents were uncertain about the safety of GM foods. This may simply imply that Chinese consumers did not possess sufficient knowledge to weigh the preponderance of information on risks or benefits associated with GM foods and then make an informed decision to consume them or not. In the second stage of the survey, however, the respondents, after being given general information on GM foods, indicated they would be more willing to buy such foods. In the third stage, the respondents were given further specific information on the merits of GM products, such as pest resistance, nutritional enhancement, and medical function, as well as on health problems found in both labora-tory and real-world settings; consequently, their acceptance increased signif-icantly, ranging from 33 percent for the pest-resistance characteristic of GM foods to 48 percent for the medical-function quality. Even the group that was given additional information on negative effects of GM foods showed higher acceptance and a more positive attitude.[37] This seems to confirm that "the more people paid attention to agricultural biotechnology in the media, the more knowledgeable they were about the issue, . . . the less they perceived risk in the technology."[38]

The Biosafety Certification for Bt Rice

As discussed in the previous pages, on August 17, 2009, after rigorous exami-nations that had taken ten and five years respectively, MOA, on the rec-ommendation of China's National Biosafety Committee of Agricultural GMOs, approved the issuing of biosafety certificates to Huahui 1 and Bt Shanyou 63, two rice strains containing Bt genes, and phytase maize. The moves were significant as they might soon open the door to the commercial growing of Bt rice, a staple crop, in China. As we also know, amid growing pressure for greater public understanding of its possible adverse impacts, the government postponed concrete measures to put forward the commercial-ization of Bt rice and allowed these biosafety certificates to expire before renewing them. Nevertheless, the prospect of commercializing Bt rice and other types of GM rice had received extensive media coverage. An analysis of such coverage, therefore, may help us understand the stimulation of tre-mendous resistance from various quarters of Chinese society and the spirit of distrust awakened among some Chinese citizens toward GMOs and the scientists involved in the research and commercialization.

While the coverage of Bt-rice biosafety certification in the two weeklies, which to a certain extent was also provoked by Greenpeace, will be dis-cussed in the next section, the reporting in *People's Daily*, *Science and Technol-ogy Daily*, and *Farmer's Daily* shows two distinct characteristics. First, there was an emphasis on the rigorous processes that both Bt rice and phytase maize had gone through before receiving their biosafety certificates. On March 3, 2010, for example, both *People's Daily* and *Farmer's Daily* carried Q&As on transgenic technology and biosafety by the MOA's Office of Agricultural Genetic Engineering Biosafety Administration. On March 15,

Science and Technology Daily ran similar Q&As. Second, the coverage emphasized that issuing biosafety certificates was not necessarily equivalent to giving the green light to commercial planting of both GM crops. The next and necessary steps are the crops' value for cultivation and use (VCU) verification, variety registration, and production license, which would take several years. Stories regarding post-biosafety-certification management also appeared in the three newspapers, giving their respective interpretations of the MOA decision. These stories were mostly about the effectiveness and necessity of the commercial cultivation of GM crops and gave only limited space to discussion of their possible adverse impacts, which were mentioned only in rebutting such impacts.

Ordinary Chinese were not prepared for the prospect of Bt rice on their dining table, which was not picked up by the mainstream newspapers under study; nor did the newspapers devote space to reporting the activism against GM crops and their scientists, as discussed in chapter 5. It was *Science*, a leading international science journal, that ran a piece by Richard Stone, its Asia editor, in its February 25, 2011 issue, detailing the views of both sides of the GM-crop debates, especially those of the activists. This piece also described an attack on the Bt-rice research carried out at Huazhong Agricultural University and Zhang Qifa, the scientist leading the effort, dwarfing the coverage of the episode in the newspapers under study.[39] On March 3 *Science and Technology Daily* published a translation of the news story, and on July 8 the same newspaper published an op-ed calling for accelerating the commercialization of GM crops and maintaining food security. Huang Dafang, one of the leading agricultural biotechnologists, started the piece with a statement that the party and government always attach great importance to and support the research and application of agricultural biotechnology. Instead of advancing his case on a scientifically grounded and evidence-based footing, Huang used the party-state stance to show the necessity of developing biotechnology and his political correctness, which is obviously unprofessional and unconvincing. Such a statement even looked strange.

Science and Technology Daily was most active in 2012, running a series, "Transgenic Technology in Other Countries," introducing the experience of developing transgenic technology and GMOs in the United States, the European Union, Brazil, India, South Korea, and Australia. These stories were mainly to dispel the rising negatives within China. Published in a daily under the nation's science bureaucracy, they also tried to convey directly government's determination to push forward the implementation of policies and measures pertaining to GM crops and foods and the application of transgenic technology. Indeed, the government had long favored transgenic technology, evidenced by the fact that the National People's Congress Standing Committee invited Huang Dafang to lecture on the technology to its members on June 25, 2010, which was reported in *People's Daily* and selected one of the top-ten events in the communication of science and technology of the year.[40] Nevertheless, given lingering concerns, from 2012 on, policy started to shift toward precaution with no concrete measures taken to promote the production of Bt rice.

Coverage in *Southern Weekend* and *Outlook Weekly*

Of the five newspapers under study, the two weeklies, *Southern Weekend* and *Outlook Weekly*, are known for their investigative journalism's sharpness and critical stands. It is worth analyzing their coverage of GMOs. Between 2000 and 2012, each published fifteen investigative news stories and commentaries around the controversy and risks associated with GMOs and the policies and measures for GMOs implemented in different countries. The coverage detailed a variety of aspects, including conflicts of interest between independent research and funding agencies, the hindrance effect of patents on research and commercialization, the redistribution of benefits brought by transgenic technology among various stakeholders, and others.

Among a seemingly critical coverage, *Southern Weekend* also carried a series of supportive commentaries, which is consistent with its more liberal style of reporting. As early as 2004, *Southern Weekend* published "GM Rice for 1.3 Billion Chinese: A Game of Safety and Interests," emphasizing many of the problems associated with GM rice, from possible adverse impacts on human health and the environment to concerns over intellectual property rights to conflicts of interest among scientists. It challenged the prospect of commercializing GM rice in China and revealed the hesitation of China's GMO policymaking.[41] The story apparently drew much of the information from Greenpeace's allegations, as noted in chapter 5.

Soon after the news that biosafety certificates had been issued to Bt rice and maize was made available in early December 2009, *Southern Weekend* published another piece, "GM Rice Allowed to Plant: Bashful Policy Released Quietly," on December 10.[42] It ran a further story on May 13, 2011, with the title "Search for the Truth of Safety of Transgene in China: The Hidden Secrets," following the tone of the paper's 2005 investigative story, "GM Rice Surprisingly Appears on the Market," on the illegal growing and sales of GM crops.[43] It is most noticeable that on July 22, 2011, *Southern Weekend* published a commentary by Ke Bei, a domestic biologist whose name is presumably not real, attacking the ignorance and prejudice toward GMOs.[44] A week later, on July 28, 2011, the weekly published a 7,000-word investigative report analyzing Chinese-style fallacies and rumors about GMOs.[45] Taken as a whole, the coverage of GMOs in *Southern Weekend* was fairly balanced during the period of the study, reporting views of both the pro- and anti-GMO camps.

The coverage in *Outlook Weekly* was more constructive, positive, and forward-looking, although it was cautious toward the commercial cultivation of GM rice. For example, in 2004, it published a story titled "GM Rice Cultivation: Why Should China Aim for the First?"[46] Five years later, on December 7, 2009, soon after biosafety certificates were issued to Bt rice, *Outlook Weekly* came up with a piece, "Cautions on Commercializing GM Rice." Then, the February 8, 2010, issue of the weekly ran a cover story, "Doubts about GM Rice," with a collection of five articles discussing in depth all the issues related to the GMO controversy in China and elsewhere.[47]

Summary and Discussion

Through analyzing the coverage of GM crops and foods in five leading Chinese newspapers between 2000 and 2012, this chapter found that the trend of coverage, measured by the number of stories, corresponded and responded to major events involving GMOs in China. Different newspapers presented or represented the views of their respective constituencies, with such themes as progress and achievements, food safety, and debate on risk leading the coverage, followed by themes of policy and governance and of technology. The coverage, at least in the five newspapers under study, was mostly promotional initially and became precautionary only after biosafety certificates were issued to the Bt rice and phytase maize in 2009.

The fact that GMO stories had yet to appear on the front page of *People's Daily*, the CCP's mouthpiece, suggests that the relevant policy might not be high on the government's agenda. Chinese care about the global development of science and technology, including transgenic technology, which explains why such stories were published on the international news pages, while their appearance on the pages of science and technology news and agriculture news further indicates that the development of GMOs in China is still largely about science rather than commercialization. The sources of information for the news were mainly members of the research community and related stakeholders, who authored most commentaries and helped construct a generally pro-GMO frame. The perspectives of GMO activism from consumers, environmental protection NGOs, and social scientists and humanists were underrepresented, consistent with the pattern of reporting on an issue favored by the government.

However, the coverage of GMOs in the five newspapers, and in the Chinese media in general, does not seem to form an effective communication modality between scientists and the public, between the government and the public, and between science and policy. In fact, the coverage may have encountered several challenges. First, while adherence to the code of professional ethics for journalism, including accuracy, objectivity, impartiality, independent investigation, logic, integrity, truthfulness, public responsibility, and reporting different perspectives, is important, if the government put constraints on covering a certain issue, these principles may become secondary. Indeed, the media's role in China differs from that in the West in that the media have to act on and cater to the government's position.[48] During the period of study, the Chinese government seemed to be ambivalent toward GMOs—in favor of moving ahead with the commercialization of GM crops but also aware of the possibility of public disapproval. In a 2001 study, the State Council's Development Research Center instructed, "The issue of GMOs is sensitive, and we should keep reports about transgenic crops strictly in check so as to prevent unsuitable news and stirring-up (*chaozuo*) reports and reduce negative impacts. We should not engage in too much media hype and sensation."[49] In 2004, the State Council issued a confidential directive to relevant state institutions, placing restrictions on discussing GMOs in the media.[50] Despite high-level instruction, in practice, according to the science journalists interviewed, the government has not banned reporting on GMOs. However, the media have to behave themselves for fear of sanctions, especially those affecting their reputation, circulation, and advertising. And journalists, especially those at the central media outlets, have to follow government guidelines.[51] Therefore, activists seldom saw their views appear in leading newspapers such as *People's Daily*, *Science and Technology Daily*, and *Farmer's Daily*, although *Southern Weekend* and *Outlook Weekly* had room for more balanced perspectives, including reasonable and nonsensational ones from activists.[52] All these factors might explain an overall tepid coverage in the newspapers under study.

Second, journalists covering GMOs and science in general have the responsibility to provide the public, not merely their peers or scientists, with sufficient information and different perspectives to help them make informed judgments.[53] In order to meet this challenge, science journalists need to keep abreast of developments in the covered topic, maintain close contact with scientists at the frontier of research, and have an open mind to a wide range of perspectives. However, some journalists may have difficulties engaging different stakeholders, some may hype uncertainty and controversy to attract readers, and a very few may publish inaccurate and unsubstantiated stories or even help distribute rumors. These, plus the complexity and sensitivity of GMOs, have generated negative views toward journalists among some Chinese scientists, who are often hesitant, if not unwilling, to be interviewed, concerned that their views may be misrepresented or distorted. For example, an agricultural economist indicated that he had not seen one journalist in five or six years.[54] However, the inaction or inability of scientists to interact with and engage the media more effectively was also partly to blame for the leadership's inaction.[55]

Third, although the analysis in this chapter indicates otherwise, Chinese science journalists as a group are perceived to be anti-GMOs. Consequently, the Chinese public have been continually exposed to a "socially constructed version" of transgenic technology in the media.[56] For example, in 2010, a widely circulated story in *International Herald Leader*, one of the newspapers of China's state Xinhua News Agency, alleged that XY335 is a variety of GM corn that caused not only mysterious disappearance of rats but also decreased numbers of pigs, as the sows were suffering miscarriages and stillbirths in some villages of Shanxi and Jilin provinces. In reporting the story, journalists ignored the simple fact that XY335 is a hybrid corn developed by Pioneer, a subsidiary of DuPont, that was approved by MOA in 2004 and planted widely in 2008 and 2009.[57] Therefore, the Center for Science Media, an organization bridging scientists and journalists, put together a *Handbook for Science Journalists Covering GMOs*, providing science journalists with information on the science, policy, and controversy of GMOs and helping them to understand the issues around the coverage.[58]

Finally, it is difficult to establish a causal relation between the nature of the media coverage of a topic and public concern about the topic, let alone a causal relation between the nature of the media coverage and policymaking.[59] However, how the media frame and cover a controversy, and how and what they decide to include in and exclude from their coverage, can influence the public's awareness and perceptions of the controversy.[60] Furthermore, as public forums, the media play a significant role in engaging evidence-based public discourse and generating political attitudes.[61] Indeed, the traditional and, increasingly, new media are becoming sources of influence, informing the public about emerging technologies, shaping their risk and benefit perceptions of the technologies,[62] and ensuring that their input is factored into policymaking on controversial issues that "have not only scientific but also social and political implications beyond the purely technological field."[63]

The public's attention to and engagement in the discussion of controversial technologies are typically event-driven, with the media playing an intermediate role in influencing policymaking. As the psychologist Paul Slovic suggested, "every form of presenting risk information is a frame that has a strong influence on the decision maker";[64] or, as lamented by the science communications scholar Matthew C. Nisbet and his colleagues, newspapers are one of the primary arenas where controversial issues garner the attention of not only the public but also interest groups and policymakers.[65] It is true that the leading Chinese newspapers under study leaned positive in their coverage of transgenic technology and GMOs, or embedded the risk discourse in that of the benefits. However, the coverage of the Nestlé case in 2003, orchestrated by Greenpeace, introduced the "transgene" concept, the issue of labeling GM foods, the "right to know," and after all, the risk and uncertainty frame of GMOs to Chinese. Coverage of the prospect of Chinese consuming GM rice also invited the public's participation in the debates. In fact, the delay in issuing biosafety certificates to GM rice in 2004 was attributed to the *Southern Weekend*'s investigative story on December 9, 2004,[66] which in turn was coincident with, or inspired by, the release of a Greenpeace report, *The Risks of Chinese GM Rice on Health and the Environment*, eight days earlier. Given the nature of media control and censorship in China, it is difficult to conclude that the media were able to influence the policy agenda directly through their coverage, although such a possibility cannot be completely dismissed. There are many other factors to consider, but the combined effects of the false media accusation in 2010 that XY335 is a variety of GM corn and the sensational coverage of the Golden Rice controversy in 2012, among others, might have been a one-two punch leading to the hesitation of China's policymakers toward the commercialization of GM rice.

CHAPTER VII

Patents and China's Bt Rice

On August 17, 2009, China's Ministry of Agriculture issued biosafety certificates for commercialization to Huazhong Agricultural University, permitting two GM-rice varieties—Bt Shanyou 63 and Huahui 1—to be grown on farmland in central China's Hubei Province for five years. According to China's Seed Law and regulations on the protection of new varieties of plants, these Bt-rice varieties still needed to undergo tests on value for cultivation and use (VCU) and to complete crop-variety registration and product licensing, which usually takes three to five years. However, this landmark decision on China's commercialization of GM crops signaled that the world's largest rice-producing and -consuming country might soon become the first to produce GM rice, a major staple crop, with the prospects for its global adoption and acceptance.[1]

The biosafety certification of Bt rice immediately spurred a heated public debate and controversy, as discussed in previous chapters. Without clear perceptions of transgenic technology, ordinary Chinese seemed to be very cautious about potential risks of GM rice to human health and the environment, as surveys conducted in such big cities as Beijing, Shanghai, and Wuhan showed.[2] Not surprisingly, the certificates expired five years later, and the commercial planting of Bt rice did not proceed as expected. The certificates were renewed in 2014 and remain valid for another five-year term until December 11, 2019.

In addition to general concerns over the effects of GM crops on human health and the environment, there were novel ones around ownership of the intellectual property (IP) in the GM rice developed in Chinese labs.[3] This chapter tries to answer the question of ownership of the Chinese GM rice through analyzing the patent portfolio of Bt Shanyou 63, one of the approved GM-rice varieties. In particular, it reviews the IP and technical property components associated with China's Bt rice with the aims of examining the distribution and structure of patent rights in the Bt rice, assessing the extent to which China owns the rights to the Bt rice, evaluating the impact of relevant patents owned by foreign holders, and identifying the possibility of patent infringements if China commercializes its GM rice.

Methodologically, this chapter is based primarily on a patent portfolio analysis by investigating the distribution of patents, particularly the coverage and strength of patent claims. The data were obtained from patent databases maintained at the United States Patent and Trademark Office (USPTO) and the European Patent Office (EPO), as well as China's State Intellectual Property Office (SIPO). A number of professional queries based on a combination of international patent classification (IPC) codes and technology-related search terms were explored to collect all relevant IP documents. In addition, relevant scientific publications are referenced wherever appropriate. The results will provide a basis for evaluating the issues associated with the commercialization of GM rice in China and reviewing China's strategy and R&D policy in agricultural biotechnology from the IP perspective. By focusing on patenting activities related to GM crops in China in general and analyzing the patent portfolio of Huazhong Agricultural University's Bt rice in particular, this chapter will reveal not only the ownership of intellectual property rights (IPRs) for the Bt rice but also the patent-infringement litigation prospect if it is commercialized and exported.

An Intellectual Property Challenge to Chinese GM Rice?

The concerns about the IPRs of Chinese GM rice were first raised by Greenpeace, the international environmental protection NGO, along with the Third World Network, another international NGO involved in issues relating to development, developing countries, and North–South affairs. Just before the biosafety certificates were granted to Bt Shanyou 63 and Huahui 1, both organizations issued major reports claiming that the Chinese-developed GM rice may fall into the "foreign patent trap," as these rice varieties use patents owned by foreign entities. In the first report, released in 2008, both NGOs claimed that at least eleven patented or proprietary methods and materials associated with three varieties of Chinese GM rice may have been previously patented by or may belong to major international agricultural business companies.[4] In 2009, in a follow-up report of a similar nature, Greenpeace and the Third World Network further concluded that another five Chinese GM-rice varieties under development might have used at least ten foreign patents.[5] Both reports implied that the Chinese GM rice would be completely vulnerable to foreign patent claims and thus fall within the scope of foreign patent protection. Greenpeace and the Third World Network further argued that the commercialization of GM rice in China would constitute a threat to China's food security and sovereignty, stable food prices, and the livelihood of small farmers. Once exported, according to the logic of the reports, these GM-rice varieties would be subject to the international IP regime, thus exposing China's rice-growing farms and the entire rice industry, including the rice-seed industry and hybrid-rice-seed industry, to the control of foreign patents. Moreover, the foreign patent holders could demand royalties and compensation from China, and the big GM-seed companies would stand to gain the largest benefits while Chinese farmers would be losers because they have no other choices, nor could they go back to conventional non-GM seeds. GM seeds are supposed to replace non-GM seeds, including conventional hybrids and inbred lines, during the extension and diffusion of GM seeds with high yield and low input. The reports by Greenpeace and the Third World Network also discussed other allegedly hidden risks associated with undisclosed and uninformed biological-material transfer agreements that the Chinese GM-rice developers may have signed with the foreign IPR holders. The GM rice under their scrutiny included the Bt varieties for which the Chinese government would grant biosafety certificates in 2009.

The possible association of Chinese GM rice with foreign-owned patents was brought to immediate high-level attention in China. However, the Chinese government and Chinese researchers involved in the development of GM rice not only disputed the conclusions reached by Greenpeace and the Third World Network, but also have repeatedly claimed that the Chinese GM-rice varieties were developed entirely using China's own materials and technology so as to allow for indigenous ownership of IPRs. For example, on July 9, 2010, MOA's Office of Agricultural Genetic Engineering Biosafety Administration declared on MOA's website that both Bt Shanyou 63 and Huahui 1 had been developed with indigenous Chinese techniques and would not infringe any patent in force filed in China.[6] This official position was then reaffirmed by a report produced by SIPO, China's patent office. The SIPO report specifically mentioned that it had granted the inventors of Huahui 1 a patent titled "The Breeding Methods of GM Rice" (Patent No: ZL200510062980.9).[7] In the meantime, some Chinese legal scholars also discounted the possibility that serious legal risks might arise during commercialization of the Chinese GM rice on the ground that China has been developing indigenous innovations and technologies. For example, of the twelve patents owned by multinational companies that Greenpeace and the Third World Network alleged might be infringed by Chinese developers, only four rights holders had applied for patents in China, and three had been granted. The first grant (CN1263946) expired on September 9, 2009. The second (CN1210402) includes claims for the gene Cry1Ab/Cry1F; but the gene used in the development of Huawei 1, which had been granted a patent from SIPO, is Cry1Ab/Cry1Ac. The third's (CN1253570C) claims also differ from the ones used in Huahui 1 in terms of the homology and equivalence of the gene sequence. That is, the claims of the second and third patents are entirely different from the techniques, methods, and materials used in the Bt rice developed by Huazhong Agricultural University.[8] The fourth application was rejected by SIPO.

In fact, there were serious deficiencies in the Greenpeace/Third World Network reports. First, these reports lacked both scientific evidence and a ground in law. Based on a search for all GM-related patents in the USPTO and EPO databases, both NGOs' research identified some patent applications with similar topics and then speculated that the Chinese GM rice would be covered by these foreign patents.[9] Second, the reports completely ignored the nature of patent protection in terms of time frame, geographical coverage, and judicial scope of the patent rights. For example, the research did not look into whether any patent at stake had been filed and especially granted in China, and if so, whether the patent was still valid. Third and most important, the Greenpeace/Third World Network research totally neglected the applicability of the foreign patent claims to the commercialization of Chinese GM rice; that is, the foreign patents cited in the reports had not been granted in any of the major rice-producing, -exporting, -importing, and -consuming countries. On the other hand, the Chinese government and Chinese legal scholars had yet to examine thoroughly whether Huazhong Agricultural University owns the full IPRs to its Bt rice. In sum, no party had obtained the full information that could shed light on the complexity and dynamism of the IP framework in biotechnology research in general, and the situation pertaining to Chinese GM rice in particular, simply because of the confidential terms of the contracts among the multiple parties involved, such as material transfer agreements and patent licensing agreements.

Research on GM Rice

Since the mid-1980s, Chinese scientists have developed many experimental GM-rice lines with such traits as higher yield; better nutritional quality; disease and insect resistance; herbicide, salt, and drought tolerance; and therapeutic functions.[10] In particular, some of these lines—including Bt Shanyou 63 and Huahui 1—have transformed genes from Bt that code for insecticidal Crystal (Cry) proteins into local rice varieties to develop insect-resistant GM rice. Scientists at public research organizations have also made efforts to establish close partnerships with private entities.[11] Research has shown that small and poor farm households in China could benefit from adopting GM rice because of both higher crop yields and reduced use of pesticides, which also contributes to reduced need for labor and improved health. For example, yields of insect-resistant GM rice are 6 to 9 percent higher than those of conventional varieties, with an 80 percent reduction in pesticide usage and pesticide's adverse health effects.[12]

As described in chapter 4, China's biosafety regulatory regime for agricultural GMOs stipulates that MOA oversee the commercialization of GM rice. Like other GM crops, GM rice must go through three phases of biosafety tests—field tests, environmental release tests, and preproduction tests—before a biosafety certificate for commercialization can be issued. Although dozens of GM-rice lines have been developed, only two—KMD1 and KMD2 transformed with a synthetic Cry1Ab gene—were evaluated in field tests in 1998.[13] As of 2005, more than a hundred GM-rice varieties and hybrids—mostly insect-resistant ones—had been in field tests.[14] It was only when GM rice was singled out as one of the priorities for the development of GM crops in the GM new variety Mega-Engineering Program, initiated under the Medium- and Long-Term Plan for the Development of Science and Technology (2006–2020), did the long-delayed biosafety approval process accelerate, which eventually led to the issuance of biosafety certificates for the Bt-rice varieties as well as for phytase maize in 2009. However, many other GM-rice lines that have also been developed and extensively tested have still been waiting for biosafety approval (table 7.1).

TABLE 7.1

Chinese GM Rice Lines Currently Approved for Field Testing or in Development

Trait	Cultivar	Developer	Regulatory Status
Pest resistance	Bt Shanyou 63	Huazhong Agricultural University	Biosafety certificate issued and expired
	Huahui 1		
	Xiushui 11	Zhejiang University	In development
	Minghui 81 and Minghui 86	CAS Institute of Genetics and Developmental Biology	
	IR 72GM Minghui 63 and Maxie 63	Huazhong Agricultural University	
	Zhongguo 91	Shandong Agricultural University	
	D297B	Sichuan Agricultural University	
	Zhuxian B	Sun Yat-sen University	
Herbicide resistance	Jingyin 119	China National Rice Institute	In development
	87203Eyi 105	Shanghai Jiaotong University	
	Xiushui 11, Qiufeng, Youfeng and Hanfeng	CAS Shanghai Institute for Biological Sciences	

Sources: Yanqing Wang and Sam Johnston, "The Status of GM Rice R&D in China," *Nature Biotechnology* 25, no. 7 (July 2007): 717–718; Hepeng Jia, "Chinese Green Light for GM Rice and Maize Prompts Outcry," *Nature Biotechnology* 28, no. 5 (May 2010): 390–391.

GM-Crop Patenting Activities in China

As a result of the state-led development of biotechnology and especially transgenic technology through public investment, as discussed in chapter 3, China has seen an explosive growth in the number of agricultural biotechnology patents. For example, in 2009, the year when biosafety certificates were issued to Bt rice, the number of agricultural biotechnology patent applications filed with SIPO was seventy-five times greater than in 1985 (figure 7.1). By the end of 2011, the total number of domestic applications for agricultural biotechnology invention patents reached 25,643, accounting for 71 percent of the total agricultural biotechnology invention patent applications received by SIPO. Compared with only one or two fillings in the early 1990s, the annual number of patent applications on GM cotton alone reached sixty-four in 2009. Of the invention patents granted, 72.5 percent

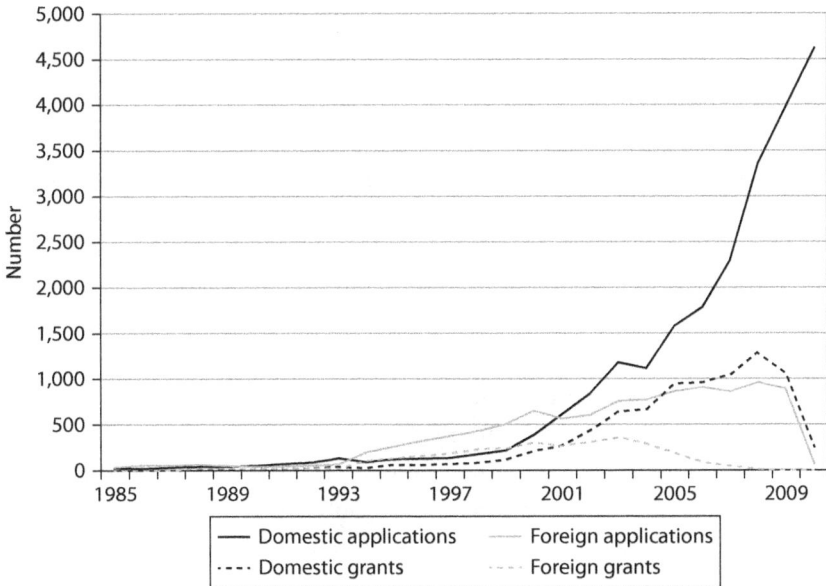

Figure 7.1 Increase in agricultural biotechnology patent applications filed with SIPO (1985–2009).

Note: The "declining" trends in grants and foreign application after 2005 result from the time lags between the filing date, publication date, and grant date of recent patents.

Source: The SIPO patent database.

went to universities and public research institutions. For example, domestic universities and R&D institutes received 266 of the 356 Bt-cotton-related patents that SIPO granted.

With IPRs in place, transgenic technology started to transform Chinese farming "from high-input and extensive cultivation to high-tech and intensive cultivation."[15] In the case of Bt cotton, Chinese scientists had successfully filed two essential patents on insecticidal Crystal protein isolated from Bt in 1995 and 1998, which can be used to produce Bt-cotton cultivars with resistance to bollworm. Now, more than 90 percent of transgenic cotton varieties planted in China has been developed using these two indigenously patented genes, which has made Bt cotton the biggest success in China.

In line with the propensity of GM-rice R&D, patent applications have also picked up remarkably. As of the end of 2011, Chinese domestic entities had filed 1,059 patents to protect a wide array of methods, materials, and products involved in the development of GM rice, which account for approximately 15 percent of worldwide applications.[16] The rising number of Chinese GM-rice patent applications paralleled the worldwide trajectory in agricultural biotechnology patenting. For example, global annual filings of agricultural biotechnology patents were no more than a hundred before 1998, but the number rapidly increased to around a thousand in 2010. However, the Chinese patent landscape shows again public-sector domination, with universities and research institutes accounting for 87.9 percent of the applications (figure 7.2). The Chinese Academy of Sciences, the Chinese Academy of Agricultural Sciences, Zhejiang University, and Huazhong Agricultural University are the main Chinese patent applicants.

There were quite few patent applications between 1985, when the GM-rice R&D program began, and 1998, when the first GM-rice field tests took place. After 2005, the number of applications surged, indicating that Chinese GM-rice R&D activities had entered a fast-growing period. However, most of the Chinese applications were filed in China only; no more than 3 percent of these patents simultaneously sought protection abroad. A patent search indicated that only one patent has been granted on Huahui 1 and Bt Shanyou 63 by SIPO and that this patent does not have family members in any foreign patent offices. In addition, the majority of Chinese patents only claim selectable markers, detection methods, and functional genes related to GM rice. In contrast

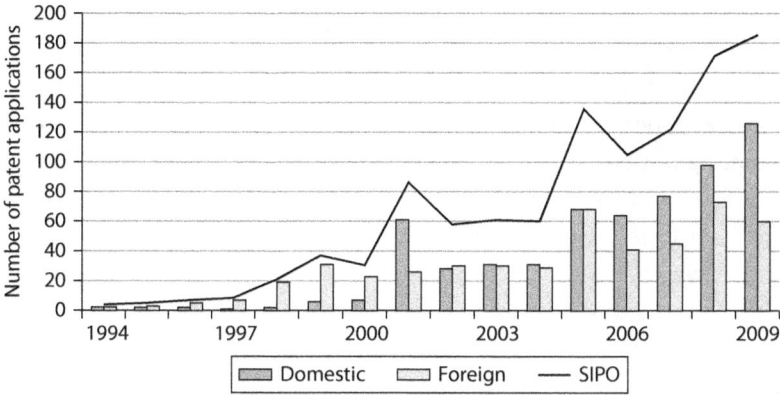

Figure 7.2 Number of GM-rice patent applications with SIPO by types of applicants (1994–2009).

Source: The SIPO patent database.

to the Chinese patenting activities that focus on domestic protection, large agricultural biotechnology companies such as Monsanto, Bayer, and Syngenta have very actively submitted their patent applications to SIPO as well as patent offices in other major rice-producing and -consuming countries. Approximately 23 percent of the GM-rice patent applications received by SIPO came from those foreign MNCs. Finally, although Chinese institutions of learning have filed more applications than foreign applicants, very few of their patents cover standard transformation methods and components used in the GM-crop R&D. This may be one of the reasons that GM opponents such as Greenpeace and the Third World Network questioned whether Chinese GM-rice lines are covered by foreign patents.

Background of Bt Shanyou 63

As discussed in chapter 3, both Bt Shanyou 63 and Huahui 1 were developed at the National Key Laboratory of Crop Genetic Improvement of Huazhong Agricultural University in Wuhan, Hubei Province. The laboratory is headed by Zhang Qifa, a well-known plant geneticist who is a CAS member (*yuanshi*) as well as a foreign associate of the U.S. National Academy of Sciences, although Zhang himself claimed that he had not been directly involved in the research. The R&D on the Bt rice receiving the biosafety certificates had been supported by multiple Chinese government funding sources, including the 863 Program and the 973 Program.

The research began in 1995. Four years later, Tu Jumin and Peng Zhongming, both Zhang's doctoral students, obtained significant results on Huahui 1 and Bt Shanyou 63, which were appraised by an evaluation expert group appointed by MOA.[17] Field tests soon began, and the environmental releases were completed in 2002. Then the final phase—preproduction testing—was carried out between 2003 and 2004. In late November 2004, both Huahui 1 and Bt Shanyou 63 were submitted to MOA's National Biosafety Committee of Agricultural GMOs for biosafety certification for commercial production.[18] However, the approval process was extraordinarily time-consuming. In fact, it took almost ten years for both Huahui 1 and Bt Shanyou 63 to go through the entire process from R&D completion to biosafety approval. Then, according to China's Seed Law and regulations on the protection of new varieties of plants, before entering commercial production, any new variety of major crops, including GM rice, must undergo tests on value for cultivation and use (VCU) and complete crop-variety registration, which usually takes two to three years and seems to be another hurdle for GM rice commercialization in China. The Seed Administration under the agricultural ministry has yet to issue a seed registration for any Bt-rice variety, nine years after issuing the first Bt-rice biosafety certificates.

As indicated, both Bt Shanyou 63 and Huahui 1 are varieties of Bt rice in which Cry genes derived from Bt conjugated with a suitable plant expression promoter and terminator are transformed, thus expressing the Bt-toxin protein to confer resistance against insects.[19] Both Bt Shanyou 63 and Huahui 1 express a Cry1Ab/Cry1Ac fusion gene, which contains a copy of the synthetic DNA sequence with two genes: Cry1Ab and Cry1Ac.[20] These genes encode the respective Bt toxins, lethal to Lepidoptera, making the plant resistant to attacks by this group of insects. Specifically, Bt Shanyou 63 is created to be resistant to rice stem borer and leaf roller.[21]

In the sequence of Bt Shanyou 63, Cry1Ab/Cry1Ac is in the middle, acting as the transformed gene of interest that is resistant to Lepidopteran insects (figure 7.3). The left side is placed with P-ract1, which is the sequence of the constitutive promoter of the rice (*Oryza sativa*) actin gene. The right side is connected with the terminator, which contains the termination sequence of the nopaline synthase gene from *Agrobacterium tumefaciens* T-DNA. The next section will analyze what technology and materials have been used

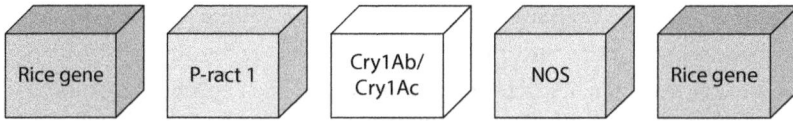

Figure 7.3 Framework of Bt Shanyou 63 composition.
Source: Biosafety Scanner software; available at http://en.biosafetyscanner.org/index.php.

throughout the entire Bt-rice R&D pipeline, which can help to achieve an informed understanding of whether the technology and materials used in the development of Chinese Bt rice are protected by any foreign patent (figure 7.4).

Two expression vectors have been transformed into the receptor material of Bt rice. One contains a Bt-gene expression cassette, and the other contains a selectable marker gene expression cassette. In the vector carrying the

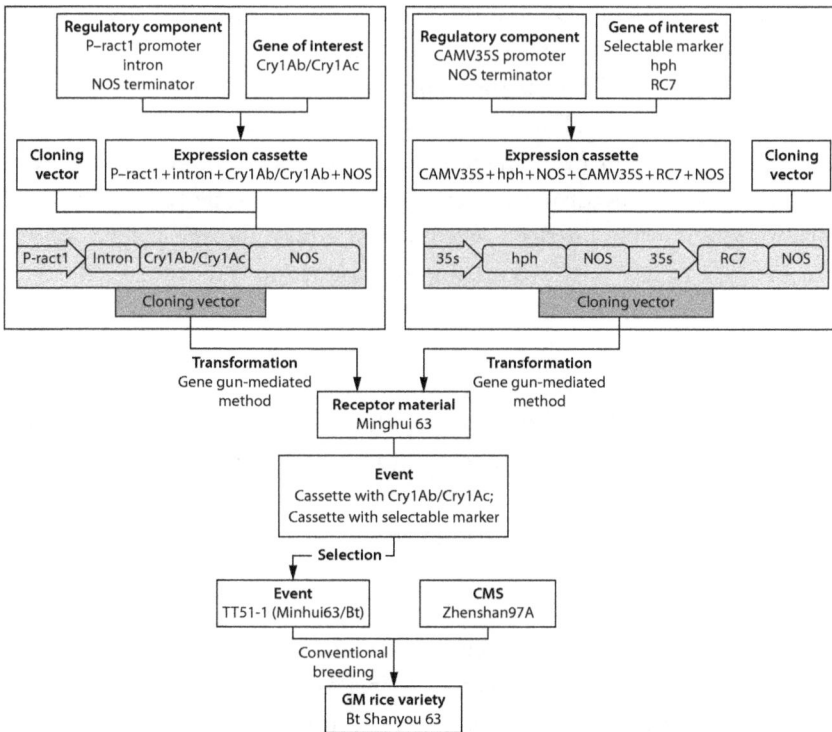

Figure 7.4 Analysis of the Bt-rice research and development pipeline.
Source: Patent document (CN200510062980.9).

Bt-gene expression cassette, Cry1Ab/Cry1Ac is integrated by the highly insect-resistance active site of Cry1Ab with the recognition site for specific insects from Cry1Ac; the regulatory element is composed of a P-ract1 promoter and nopaline synthase (NOS) terminator. In the vector carrying the selectable marker gene expression cassette, the selectable marker is Hygromycin-B-Phosphotransferase (hph) and the correspondent regulatory element consists of the Cauliflower Mosaic Virus (CAMV35S) promoter and NOS terminator. In addition, the vector contains another expression cassette, which consists of a rice chitinase gene RC7, as well as the CAMV35S promoter and NOS terminator. The two cassettes are constructed together in the same cloning vector to construct an expression vector with a selectable marker.

The two expression vectors are transformed into the reception material from a conventional rice variety—Minghui 63—using the gene gun-mediated method. Then the event with the functional Bt-gene expression cassette is kept through genetic recombination, isolation, and selection over multiple generations, and the event with the selectable marker hph is discarded. The target event is named TT51-1. By using conventional breeding method, TT51-1 is then combined with a conventional variety of rice, Zhenshan 97A. Finally, the two transgenic rice lines expressing Bt gene—Huahui 1 and Bt Shanyou 63—are generated. In fact, Bt Shanyou 63 is a hybrid of Zhenshan 97A (cytoplasmic male sterile [CMS] line) and Huahui 1, a restorer for the CMS line. Because only one hybrid variety can be used in the final production, the following discussions will focus on Bt Shanyou 63.

Patent Portfolio Analysis: Deconstruction of the Components of Chinese Bt Rice

An Overview

The deconstruction of the IP used to develop Bt Shanyou 63 is complex. Theoretically, three types of techniques and components are involved in its R&D: public domain knowledge, proprietary techniques, and technical properties. Our IP analysis is based on the first two types, which can be found in the relevant literature, including the USPTO, EPO and SIPO patent databases as well as other references in scientific databases such as the

Web of Science and China National Knowledge Infrastructure (CNKI), a Chinese academic journal database. A limitation of the analysis is that we do not have access to private information such as trade secrets, contracts, and material transfer agreements such as those between Huazhong Agricultural University, the developer, and relevant IP holders. Therefore, technical properties such as computer software, germplasm, and biological materials and derivatives are not available for examination. We will further elaborate on that issue later in the chapter.

Of the six patents that are potentially implied in the R&D of Bt Shanyou 63, five—the essential components and techniques including the promoter, transformation method, and selectable marker used in the product—are owned by foreign entities (table 7.2). In particular, the CAMV35S promoter and the gene gun-mediated transformation method are owned by Monsanto, the selectable marker hph by Syngenta, the P-ract1 promoter by Cornell University, and the RC7 gene by two Japanese public research institutes. The sixth, as the only successful event, is protected by the developer's patent in China. Additionally, although Huazhong Agricultural University submitted the application for plant-variety protection for the restorer line of Bt Shanyou 63 to MOA in early 2000, a plant-variety certificate has yet to be issued.

The Scope of Protection of Foreign Patents

In order to determine whether the Chinese Bt rice is fully covered by the patents mentioned above, we further scrutinized the scope of protection of these patents, which is defined by the claims presented in the patent documents as well as the extended rights established by national patent laws and related regulations. The interpretation of claims in the patents dealing with genes and transformation processes can be extremely complex, with certain patents having claims that appear to cover a wide range of live modified organisms (LMOs).

We have carried out in-depth analysis of the scope of protection of the five foreign patents (table 7.3). As P-ract1, CAMV35S, hph, and RC7 belong to the upstream techniques, use of any expression cassette, vector, event, or seed without obtaining licenses from the patent holders may constitute patent infringement. Likewise, if any organism is transformed with the so-called gene gun (or gene gun-mediated method), the patent protection also covers

TABLE 7.2

Patents Related to Bt Shanyou 63

Components and Methods	Title of Patent (Country and Patent Number)	Patent Holder
Promoter (CAMV35S)	Method for enhanced expression of a DNA sequence of interest (US5424200)	Monsanto
Transformation method (gene gun–mediated method)	Method of creation of transformed rice plant (US6288312)	Monsanto
Promoter (P-ract1)	Rice actin gene and promoter (US5641876)	Cornell Research Foundation
Selectable marker (hph)	Selectable marker for development of vectors and transformation systems in plants (US5668298)	Syngenta
Gene (RC7)	Complementary DNA for rice chitinase having lytic activity against molds and bacteria, and vector containing said complementary DNA and transformant (US6124126)	National Food Research Institute, Bio-oriented Technology Research Advancement Institution, Japan
Event (TT51-1)	Method of breeding genetically transformed rice (CN200510062980.9)	Huazhong Agricultural University, Zhejiang University

Sources: USPTO, EPO, and SIPO patent databases.

TABLE 7.3
Scope of Protection of Foreign Patents

Stage	R&D					Commercialization	
Coverage	Amino Acid	Nucleic Acid	Expression Cassette*	Event		Seed	Nonliving Organism
hph							Undefined
RC7							Undefined
P-ract1							Undefined
CAMV35S							Undefined
Gene-gun method							Undefined

*Including expression vectors carrying the cassette.
Sources: USPTO, EPO, and SIPO patent databases.

all of the events and related seeds derived from the organism. Additionally, we could not find any particular patent application on the NOS terminator when searching patent databases as far back as 1982.[22] In any event, if any live modified organism generated by Bt Shanyou 63 enters a country where these foreign patents are in force, it will definitely fall into the "foreign patent trap," according to present patent international governance, primarily under the Paris Convention for the Protection of Industry Property and the Patent Cooperation Treaty (PCT). However, patent laws in most countries do not clearly define whether patents apply to nonlive products (e.g., foods and feeds) that use a patented component or method. To date, only the European Union has issued a directive that its patent law cannot be used to bar imports of products made from biological ingredients that are patented in the EU but not in the exporting country.[23] To understand thoroughly the IP risks facing the commercialization of Bt Shanyou 63, we further investigated in which countries these patents have been filed and whether they still are in effect.

The Protection Period and Geographical Coverage of Foreign Patents

Given the nature of patent rights, the protection conferred by a patent is limited to a specified term (generally twenty years from the filing date of the patent application) and geographical area. We have surveyed the geographical regions where the relevant patents have been filed and summarized the expected term of these patents (table 7.4). All five patents in question have been granted in the United States but have not been filed in China. Meanwhile, RC7 is protected in Japan as well; and the gene-gun method has been patented in the EU, Australia, and Japan. Of these five patents, three have expired; the one on the RC7 gene has the longest remaining term of protection; it is set to expire in October 2018.

Our patent analysis confirms some of the findings by Greenpeace and the Third World Network. That is, Bt Shanyou 63 does not hold an entire patent portfolio spanning from the upstream techniques and components to final products. All genes, promoters, markers, and others used in its R&D were not developed by its developer, who simply completed the transformation and obtained the target event. But the event has been protected by one Chinese patent granted to the developer.

Table 7.4

Remaining Term of Patent Protection in Target Regions

Components and Methods	Target Region	Remaining Term								
		2012	2013	2014	2015	2016	2017	2018	2019	2020
hph	US				Sept. 15, 2014					
RC7	US, JP									Oct. 25, 2018
P-ract1	US				June 23, 2014					
CAMV35S	US		June12, 2012							
Gene–gun method	US, EU, AUS, JP									Sept. 10, 2018

Note: The dates indicate the (expected) expiration of patents in the United States; AUS = Australia, JP = Japan.

Sources: USPTO, EPO, and SIPO patent databases.

However, the Greenpeace/Third World Network reports have obviously exaggerated the challenges facing Bt Shanyou 63. First, they diverged from those literally relevant patents when breaking down the pipeline of the Bt-rice R&D. Second, they did not identify whether these foreign patents had been filed—or, more importantly, granted—in China. For example, the reports argued that the four Bt-gene patents held by foreign entities may cover the Chinese Bt rice. After examining these patents, we have found that they are about the fusion of Cry1Ab/Cry1Ac and Cry1F, Cry2, and VIP, but their gene sequences are completely different from those of the gene of interest transformed in the Bt Shanyou 63. So the Greenpeace/Third World Network reports have a major flaw: they just identified patents with similar titles without looking closely into their technical details. Except for P-ract1 and CAMV35S, other patents actually have nothing to do with the development of the Chinese Bt Shanyou 63. By stating such a blanket conclusion, whether intentionally or unintentionally, Greenpeace and the Third World Network confused and indeed misled the Chinese public and probably the political leadership as well. Finally, the Chinese developer does own a patent on the event, which was granted by SIPO, China's patent office.

Other Potential IP Risks

If only patent protection is taken into account, the production of GM rice, Bt Shanyou 63, within China appears to be exempt from foreign patents because their owners have not sought patent protection in China. However, in addition to patents, contracts and trade secrets are widely considered part of the IP; in particular, material transfer agreements signed between a provider and a recipient are not limited to a specific geographical area. With these contracts or agreements, which likely exist but are not in the public domain, the IP risks facing the Chinese Bt rice could be more complex, because the use of components and methods is subject to specific contract clauses in these agreements, even though these components and methods are not patented in China. Additionally, the components used in the Bt-rice R&D can be defined as genetic resources; therefore, their utilization is further governed by the Nagoya Protocol on Access to Genetic Resources and Benefit-Sharing.[24]

Along with the implementation of the Nagoya Protocol in the member countries of the Convention on Biological Diversity (CBD), of which

China is a signatory, the parties that commercialize Bt Shanyou 63 may also face demands for sharing the benefits by the owners of relevant genetic components (e.g., promoters and selectable markers). However, because proposals on mandatory disclosure of the sources of genetic resources in the patent application might not have been accepted by the Agreement on the Trade Related Intellectual Property Rights (TRIPs), and the enforcement of Nagoya Protocol is mainly subject to national legislations in the contracting parties, benefit-sharing would not be so burdensome for the commercialization of Bt Shanyou 63 in the near future.

Summary and Discussion

Clearly, the question of who owns the IPRs to the Chinese GM rice is quite complex because the GM-rice R&D is a long-chain process, including but not limited to many sophisticated components and methods from genes of interest, promoters, terminators, selectable makers, expression cassettes, expression vectors, and transformation methods. In the process, the Chinese developer at Huazhong Agricultural University may, intentionally or unintentionally, have used others' proprietary technology, including patent rights, plant-variety rights, trademarks, trade secrets, know-how, and germplasm and other biological materials. Through deconstructing the process of development of the Chinese GM rice, Bt Shanyou 63, this chapter has reviewed reception materials, gene constructs of cloning vectors, transformation, plant regeneration, and other related techniques.

The analysis indicates that the R&D of Bt Shanyou 63 is potentially covered by approximately five foreign patented components and technologies—fewer than the Greenpeace/Third World Network estimate of a dozen or so foreign patents. Furthermore, it must be clearly stated that because none of these patents have earned protection in China; if there were no relevant material-transfer agreements or other contracts signed between the Chinese developer and foreign patent owners, there would not be any IP risk associated with commercial growing of the rice varieties in China. Therefore, the argument by Greenpeace and the Third World Network is not valid, and Chinese farmers do not have to pay royalties to foreign companies and organizations for growing Bt rice in China.

However, it also is premature to conclude that there would be no challenges facing China in commercializing the Bt rice in the international market. In particular, if the seeds or other live modified organisms (LMOs) generated from Bt Shanyou 63 flow into the United States, the EU, Japan, or Australia before late 2018 while some of the above-mentioned patents are still under protection in these countries, the Chinese developer could be sued by the relevant patent holders. Moreover, given that there are no clear internationally accepted rules on whether nonlive products are subject to biotechnological patent protection, trade in foods and feeds produced with the Chinese Bt rice may also face lawsuits in these countries.

While the analysis in this chapter only covers Bt Shanyou 63, its findings should not be limited to this particular variety of the Chinese GM rice and can have implications for other GM-rice varieties or even GM crops developed in China as a whole. In retrospect, the case of Bt cotton in China may present a lesson. As discussed in chapter 3, Bt cotton is the biggest success of the Chinese agricultural biotechnology program. China started to grow Bt cotton in 1997 and has seen its share increase to more than 90 percent of all cotton planted in northern China since 2004.[25] On one hand, Chinese Bt cotton has faced a similar international patenting landscape to the one that Bt rice is facing. Foreign companies such as Monsanto have not patented their Bt genes in China, although they own patents on these genes that are important in the production of the cotton, including that developed by Chinese agricultural biotechnologists and companies. In fact, until recently, MNCs producing GM seeds have been impeded from carrying out related R&D in China, as alluded to earlier, which may affect their patenting activities in China. They have also taken an approach of not bringing their more sophisticated and advanced seeds into China because of concerns about the country's weak IP protection. Nevertheless, Bt-cotton commercialization has been done without the risk of patent infringement, at least in the domestic market. The Bt-cotton case could free the Chinese government and developers from concerns about potential patent lawsuits relating to GM rice. Thus, the government, and especially ambitious developers, could confidently distribute Bt Shanyou 63 on a large scale without facing challenges on the patent front.

On the other hand, the developer of Chinese Bt cotton—the CAAS Institute of Biotechnology—has established a more rigorous and efficient IP protection framework than its Bt-rice counterparts. For example, the institute has patented the Bt genes inserted into all of the Bt-cotton varieties produced in China. A transformation process for inserting genes— the pollen-tube pathway system—was developed by Zhou Guangyu and her Chinese colleagues.[26] And most of the Bt-cotton varieties have been covered by plant-variety rights. By contrast, Bt Shanyou 63 has not been granted a plant-variety right more than fourteen years after application, and no substantial advances in the progress of commercialization have been made since the biosafety certificates were issued and renewed. Thus, if it is to be commercialized in the domestic market, the developer and seed companies cannot ensure economic benefits comparable to those of Bt cotton. And the market incentives for Chinese seed firms to commercialize Bt rice will be insufficient.

All in all, from a practical perspective, the lessons from both Bt cotton and Bt rice should convince the Chinese government and GM-crop developers of the necessity to have a dynamic IP risk-management strategy. These lessons, in turn, should enable Chinese transgenic scientists and developers to refine their R&D strategy. In order to avoid infringement, they need to perform IP analysis *before* starting research rather than at the stage of commercialization, which is too late. Meanwhile, they must also take preemptive measures to protect their innovation.

As stated, the IP analysis of Bt Shanyou 63 relies on the patent documents and the scientific literature. The information on nucleic-acid and amino-acid sequences of the GM-rice components disclosed in some patents is incomplete, and the details of technical properties such as material-transfer agreements are not available for examination. Had we had such information, we would have further examined its impact on the commercialization of Chinese GM rice.

Furthermore, the analysis applies only to the ownership of IPRs of Bt Shanyou 63 and Huahui 1, not to other varieties of GM rice that have been developed. For example, the gene used to develop rice resistance to bacterial leaf blight by Jia Shirong at the CAAS Institute of Biotechnology was secured from the University of California, Davis, through Jia's partner at the International Laboratory for Tropical Agricultural Biotechnology, the Scripps Research Institute.[27] Other well-known Bt-rice varieties—KMD1 and KMD2—under attack by Greenpeace and the Third World Network have been developed by the Institute of Agricultural Nuclear Science at Zhejiang University jointly with the University of Ottawa.[28] The extent to which these parties have agreed on IPRs and profit-sharing is not available for examination. Nevertheless, concerns over the ownership of IPRs of GM rice developed in China have inevitably become a factor in the decision by the Chinese government whether to promote the commercialization of

GM rice. This may also explain why Greenpeace attributed the inclusion of an emphasis on indigenous IPRs in Central Document No. 1 in 2010 to the reports that it and the Third World Network issued on the foreign patent ownership of Chinese GM rice.[29]

Finally, the overall IPR protection involved in the R&D of GM crops is broader than the issues brought by Greenpeace and the Third World Network on GM rice. We have seen a similar rationale—the protection of domestic soybean farmers—coming into play in China's establishment of its biosafety regulatory regime. Indeed, China is facing a dilemma in addressing the conundrum of food security—whether to increase production even if some of the IPRs are foreign-owned or to accept lower production in the hope that eventually domestic IPRs will render foreign IPRs unnecessary.

CHAPTER VIII

China as a GMO Nation

The documentary on GMOs in the United States that Cui Yongyuan shot and put on Chinese internet portals (see this book's introduction) received millions of hits. Cui also submitted his proposals regarding illegal planting of GM crops to the Chinese People's Political Consultative Conference as promised. While not evoking such assertions as "GM crops are a conspiracy of American imperialism to annihilate Chinese" or "Chinese GM scientists are traitors," the documentary did touch some elements of the Chinese psyche, even morality and patriotism, if not nationalism. The GMO activism represented by Cui Yongyuan, Greenpeace, and conspiracy theorists, as discussed in this book, has gathered momentum and negatively affected the trajectory of China's GMO policy. The fact that the biosafety certificates issued to Bt rice and phytase maize were allowed to expire in August 2014 was a significant setback to Chinese as well as global efforts toward commercializing GM crops. This concluding chapter will summarize findings of the book and extend the discussion to an analysis of the implications for policymaking of post-academic science and post-normal science as well as S&T policy culture in China's ongoing pursuit of transgenic technology.

Intertwined and Multifaceted Factors

The book has discussed China's evolving policy toward GMOs along several policy dimensions: public research investment and efforts, the establishment and strengthening of a biosafety regime, the rise of GMO activism that also has involved food safety and consumer choice and trade, and intellectual property rights (IPRs).[1] These policy dimensions either are embedded in or overlap with the interests of various stakeholders.

Scientists and Business

The development of transgenic technology in China has mostly taken place in a promotional policy environment, at least on the research side. With strong government support, Chinese scientists were among the first in the world to research GM crops, playing a pioneering role, individually or collectively, in advancing scientific knowledge. In 1997, the government approved commercialization of Bt cotton, allowing the country to embark on a path apart from other developing countries such as India and Brazil. Facing uncertainties and other factors, Chinese scientists have tried to confine the debates to the domain of science.

Contrary to the situation in the United States, where business interests have been central in shaping the policy over transgenic technology and GMOs,[2] in China, scientists have actively set the policy agenda. In doing so, they have become an interest group, tied to other powerful interest groups, especially the government. Encouraged by the mandate of the reform of the S&T system to improve linkages between research and the economy, some Chinese scientists, including transgenic scientists, have become technological entrepreneurs themselves, engaging in the commercialization of their research. This has been a very positive and encouraging development.

While participation by domestic business in agricultural GMOs still is limited, the interests of MNCs have been evolving. For example, among the first to bring transgenic technology to China, Monsanto formed joint ventures with Chinese seed companies and collaborated with Chinese institutions of learning such as Huazhong Agricultural University. MNCs also have been able to influence China's GMO policy. As discussed in chapter 3, the virus-resistant tobacco was removed from production simply because of pressure from the U.S. Department of Agriculture on behalf of an American tobacco maker. However, China's agricultural biotechnologists have been scrutinized over whether they have advocated a promotional policy for the sake of commercializing their own research or simply catering to the interests of MNCs. They have even been challenged as to whether they would have acted differently had they not had a stake in the companies that they helped to set up for the commercialization of GMOs. Mistrust and cynicism toward science and scientists have been reinforced as some of the scientists do have conflicts of interest as both academic scientists and entrepreneurs.

Tensions have also built up between transgenic and nontransgenic scientists. Apparently, the allocation of significant public funding to research on GM crops has put scientists working on conventional breeding, organic agriculture, and ecological farming in a disadvantageous position.[3] The Medium- and Long-Term Plan for the Development of Science and Technology's channeling of some RMB20 billion into a mega-engineering program for new GMO varieties has been further seen by nontransgenic scientists as ignoring their lines of work. Thus, nontransgenic scientists' objections to promotional policies for GMOs may reflect dissatisfaction and resentment over their own diminished professional and career prospects.

NGOs, the Media, and the Public

Transgenic technology is one of the technologies perceived to have uncertain impacts on health and the environment, thus leading to rational or irrational concerns and sounding necessary or unnecessary alarm among the public. Given this, consumers have sufficient reasons to ask for a "right to know" (*zhiqingquan*) about whether food products on the market are genetically modified, and if so what the GM ingredients are, to inform their purchase decisions. Philosophically, the "right to know" cannot be ignored or neglected, and the public's involvement in the policymaking process is warranted.

China's domestic NGOs have advanced their interests in environment-related issues[4] but have largely been missing in the debates on GMOs.[5] This has given Greenpeace, the international NGO and an unlikely actor in China's domestic affairs, an opportunity not only to participate actively but also to become a major anti-GMO force and even influence regulations and policy. Early in this century, when China's State Council and Ministry of Agriculture were formulating regulations and measures on agricultural

GMOs, Greenpeace did more than just rhetorically oppose the commercialization of GM crops by disseminating misinformation on GM products;[6] it took concrete actions. It challenged Nestlé on the labeling of foods containing GM ingredients and launched a "right of choice" campaign to raise awareness among Chinese consumers of the safety of GM foods. It became de facto police in China's dairy and later other food markets by identifying, examining, and blacklisting products with GM ingredients, thus stirring public fear toward transgenic technology as opposed to other new technologies. It often reminds the Chinese public that transgenic scientists have dual identities as both researchers and shareholders in companies that utilize their technology, suggesting that they have promoted the commercialization of GM crops purely for their own interests. Thanks to Greenpeace, transgene (*zhuanjiyin*) has been turned from a novel and unfamiliar scientific concept in the late 1990s to a topic that ordinary Chinese seem capable of talking about in the first decade of the twenty-first century. However, it is difficult to understand the Chinese government's tolerance of this international environmental NGO, given its early tensions with Greenpeace on nuclear testing as well as its overall suspicion toward international NGOs.

Uncertainties around GMOs obligate the media to provide balanced information and views to help the public understand them. Chinese media coverage of GMOs has witnessed a dramatic increase since 2000, and transgenic technology and GM foods have been widely discussed and debated. In order to attract attention, some Chinese newspapers have published stories that have been proven inaccurate or false. In addition, the internet and social media have increasingly become a public space where not only have concerns over GMOs been actively raised but also misinformation and disinformation on GMOs, GM scientists, and relevant government officials have been disseminated. As "the links between media exposure, levels of awareness, and attitudes toward biotechnologies are far from straightforward,"[7] and the relationship between the "actual inputs" of the media and "the [policymaking] process or the specifics of the process itself" is unclear,[8] it is premature to conclude definitively that such coverage has affected—either positively or negatively—the evolution of China's policy on GM crops and foods. Nonetheless, media coverage has at least helped to gradually shape a precautionary attitude toward GM crops among the public,[9] not only making GMOs a topic beyond science but also forcing policymakers to pay attention to different views and to take the interests of various stakeholders, especially consumers, into consideration in fine-tuning the policy.

Regulations

Along with the involvement of NGOs, the media, and the public in the debates on GMOs, China has witnessed the establishment and enhancement of a biosafety regulatory regime of agricultural GMOs. The emergence of such regulations was not merely to meet the challenge from a precautionary stand on GM foods in some parts of the world, especially in Europe. Domestic concerns have also prompted China to regulate GMOs to balance the rapid advancement at the front of science and especially the commercialization of transgenic science.[10]

Until the late 1990s, the biosafety of agricultural GMOs was a matter of science in China, confined to the safety of genetic-engineering research, experiment, and production. This explains why the initial regulator of agricultural biotechnology was the State Science and Technology Commission, a science bureaucracy. Responsibility for managing biosafety was later transferred to the Ministry of Agriculture, an agricultural-production-oriented bureaucracy. As the clearinghouse and gatekeeper, MOA set up a genetic-engineering biosafety committee to take charge of the supervision and evaluation of the biosafety of GMOs and the administration of the mandatory approval for field testing, environmental releases, and commercialization of GM crops. In 2001, the State Council stepped in to prescribe an official procedure for the approval of all production and marketing of GMOs and the certification of imported GM products as harmless to consumers, animals, and the environment before they can be sold on the Chinese market. The MOA measures to implement the State Council regulations that followed further required mandatory labeling on foods containing GM ingredients. The policy was meant to shift biosafety and food-safety regulations from a product focus, as in the United States, to a process focus, as well as to impose what has been referred to as the world's most restrictive regulations and measures on the research, production, retailing, and import of GM crops, similar to those in Europe, thus frustrating scientists specializing in agricultural GMOs.

While European policy has put greater weight on the environment, China's increasingly stringent biosafety regulations and related policy discourses have given the agricultural ministry regulatory primacy on GM crops. The State Environmental Protection Administration (SEPA) and its successor, the Ministry of Environmental Protection (MOEP) and the

Ministry of Ecology and Environment, have played a limited role by serving merely as a member of the Joint Ministerial Conference that the agricultural ministry convenes. However, given its representation of China in the international negotiation of the Convention on Biological Diversity and the Cartagena Protocol on Biosafety and its role in drafting China's national biosafety framework, the ecology-and-environmental-protection apparatus may not be willing to play only a secondary role. SEPA formed a partnership with Greenpeace as early as 2002 to examine whether the cultivation of Bt cotton caused adverse environmental impacts. In 2011, MOEP was instrumental in the first large-scale investigation of the illegal planting of GM crops.[11] It has also been noticeably campaigning for the formulation of a biosafety law, which, if enacted, would have a legal status higher than the State Council regulations and MOA measures.

In recent years, regulations on the research on and commercialization of GMOs have not been stringently implemented. For example, Greenpeace has on multiple occasions uncovered unauthorized planting and sales of Bt rice in Hubei Province, and ultimately forced the provincial government to destroy such crops. They have also raised concerns about collusion of scientists and regulators in activities violating regulations and whether scientists have confined their work within the framework of existing laws and regulations and research governance. The trial of Golden Rice among Chinese school-age children without proper regulatory and research ethics approval and monitoring provided further grounds for Greenpeace and China's domestic GMO skeptics to advance their causes. Unfortunately, these cases have not only prompted a public outcry but also tarnished the image of transgenic scientists and research and commercialization efforts as a whole.

Trade

While the precautionary and preventive aspects of China's GMO policy resonated well with the global uneasiness toward GMOs' possible environmental and health effects, practical international trade considerations have been facing China as well.[12] The year 2001, when China rolled out its biosafety regulatory regime, coincided with its trade disputes with the United States, Brazil, and Argentina on GM soybean imports. China's biosafety regulations were thus perceived as the basis for opposing the imports of GM soybeans and protecting China's less competitive domestic soybean farmers. Upon becoming a member of the World Trade Organization in late 2001, China was no longer able to use tariffs and nontariff barriers to limit GM soybean imports. Thus, biosafety regulations seemed to be a rational choice, although under WTO rules overly long and complicated transgenic event-approval procedures are considered as trade barriers.

The trade issue was further complicated by the fear that China could lose its advantages in agricultural exports. Restraints on trade in GMOs based on sanitary and phytosanitary grounds, which are allowed under WTO, have increased around the world. Chinese rice has been repeatedly singled out for contamination with GM ingredients. Therefore, by limiting the varieties of agricultural GMOs being grown, China may also want to promote its food and agricultural exports as GM-free and retain its export markets in Europe and Japan.[13] In this regard, China's contradictory policy of encouraging research on GMOs but restricting the commercialization of such crops has put the nation in a position similar to that of the United States, which, as the world's largest producer of GM crops, has to restrain itself from exporting such products to countries that ban or have tight regulations on them.

However, China has to meet its increasing demand for soybeans, corn, and other agricultural commodities. The rise of the newly affluent middle class and the easing of the one-child policy will lead to skyrocketing demand for meat, eggs, and edible oil, as well as for corn and soybeans used in their production. Unable to produce enough domestically, China has to source these crops from the international market, thus providing opportunities for countries that export such commodities, many of which are genetically engineered. For example, China imported more than 80 million metric tons of soybeans and 2.63 million tons of corn, most of which are GM varieties, in 2014. However, in 2013 and 2014, China also turned away GM-corn shipments from the United States because they were found to contain MIR162, a variety developed by Syngenta that China had not approved for import.[14] This move cannot be simply construed as a trade barrier, as biosafety certificates for ten other imported GM-corn varieties from Monsanto, Syngenta, and DuPont Pioneer were still valid at that time.[15] (MIR162 was approved for import in late 2014, and other GM varieties have been approved since then.[16])

Key Issues of the GMO Debates in China

Globally, the debates on GMOs have been about the labeling of GM foods, the possible environmental effects of GM crops such as gene flow and resistant superbugs and superweeds, the respective interests of farmers and consumers, and alleviating and resolving food-security problems. Other related issues include the objectivity of scientific research and publication, the role of regulations, and public participation, among others. In China, the debates have centered around three closely associated issues around, beyond, and related to GM crops and foods—food safety and biosafety; foods security and sovereignty; and policymaking.

Food Safety and Biosafety

Answers to questions of food safety and biosafety are in the domain of science, mainly depending upon experts and knowledge. The mainstream global scientific community recognizes that GM crops and foods approved by regulators in respective economies do not pose adverse impacts on human health and the environment. As most GM crops have pest-resistant traits, agricultural biotechnologists argue, planting such crops can also reduce the application of pesticides, thus countering the view that stresses the unknown, potential risks of GM crops while ignoring the known, real, direct, and even fatal consequences of chemical pesticides. However, the science- and evidence-based discussion on GM-food safety and biosafety has not been fully appreciated in China. The anti-GMO activism directly challenged it, claiming that it is still too early and too soon to reach such a categorical conclusion. It also evoked the historical case of DDT and *Silent Spring*, the book that documented the pesticide's detrimental effects on the environment, to insist on the necessity of a longer period of empirical study that should also involve human beings.

Food Security and Sovereignty

GMO activists have escalated the debates over the issue of food security and sovereignty politically and ideologically. Greenpeace is always critical of the domination or monopolization of GM patents by a small number of MNCs. This is also one of the aspects that Chinese activists have used to undermine the commercialization of GM rice. Once large-scale cultivation of GM rice commenced, they contend, China would not only have to pay high royalties but also concede its food production to foreign patent holders. Moreover, according to Peng Guangqian, a major general in the People's Liberation Army and deputy secretary-general of the National Security Forum, GM foods not only cause cancer and are a source of infertility; they have also been created by Monsanto, an American firm with a notorious history, backed by the Pentagon and leading American private foundations, to control the global food chain and threaten China's food security and food sovereignty.[17]

The views expressed by Major General Peng are more than the public expression of the conspiracy theory targeting GMOs. Indeed, the use of sensational language to attack GMOs and their creators is all over the Chinese internet and social media. In the view of the activists, Monsanto is a biological weapon that the United States uses against China; the introduction of GM crops to China is no different from the United States launching a "new Opium War" targeting China; GMOs are part of a grand Western strategy and a monumental, supremely devious plot to annihilate Chinese and other people of color; and Chinese transgenic scientists who have studied GMOs and government officials who have given the green light to GMOs are traitors of the twenty-first century. Such language clearly indicates that the debates are a war of words rather than reason.[18] It may indicate that Chinese society has become pluralistic, allowing open expression of different opinions, but the anti-GMO sentiment vents anti-Western and anti-American emotions, rather than rationally discussing science and its possible impacts on society.[19] It is also beyond comprehension that the promotion of such anti-government-policy views and defamation of scientists and government officials have been tolerated on the Chinese internet, where tight censorship continuously scrubs criticism of government policy and silences dissents.

However, Chinese GMO activists have missed a crucial point. That is, it is precisely the concerns about food security and national peril that make it necessary for Chinese scientists to develop GM crops indigenously while absorbing advanced technology from abroad. In addition, China has never had its doors wide open to transgenic seeds developed by Monsanto and other companies. Only in early 2017 did China's *Catalogue Guiding*

Foreign Investment in Industries consider removing "production of geneti-cally modified seeds" from the prohibited category and confirm that the removal of the term "research and development" for GM products means that companies can obtain local cultivation permits for imported agricul-tural GM products.[20] Until now, Monsanto has confined its Chinese busi-ness to conventional breeding and has had almost nothing to do with GM seeds. That Monsanto has not sought patent protection for its transgenic technology in China is a strategic failure on its own part, just as it has not applied for patents for GM soybeans in Argentina, which also explains why Monsanto has not gained much in the GM-cotton business in China.[21] Moreover, Monsanto and its competitors in the GM-seed business have not devoted much effort to GM rice, as it is not a major Western staple crop, except in the case of Golden Rice, which Syngenta developed but has relinquished its IPRs for humanitarian reasons. The United States has never grown GM rice out of concern that GM rice would jeopardize its agri-cultural exports, although it deregulated Bayer CropScience's herbicide-tolerant rice in 1999.[22] And the LibertyLink rice contamination in 2006, discussed in chapter 1, signaled a further alarm. On the other side of the Pacific, Bt Shanyou 63 and Huahui 1, which received biosafety certificates in 2009, were developed by Chinese scientists in their own labs. In the process, Chinese scientists have generated and will continue to generate indigenous IPRs. If commercialized, the Chinese varieties will take over the domestic market and help safeguard China's food security, although it is debatable whether a nation's ownership of IPRs is determinative of the nation's food security.

Policymaking of Post-Academic Science and Post-Normal Science

The debates on GMOs in China are related to policymaking regarding transgenic technology or emerging technologies as a whole. International experience suggests the need "to open up the policy processes surrounding new technologies to far greater interaction with members of the public and their diverse values . . . while maintaining a central place for scientific information and analysis as it emerges."[23] As there is a confluence of non-scientific factors under the rubrics of the impacts of science on society and of the risks inherent in modern society itself, a purely scientific approach may not be suitable for GMO policymaking. Meanwhile, the increasing politicization of transgenic technology has further intensified the need for a more inclusive and democratic mode of science policymaking. In fact, making policy on GMOs and new technologies in general should involve more than academic science and normal science.

Academic science is derived from the assumption of the prominent American sociologist Robert K. Merton proposed in the 1940s that members of the scientific community are obliged to follow a set of "institutional imperatives," or norms—communism in the sense of communalism, universalism, disinterestedness, and organized skepticism.[24] Abiding by such norms, academic scientists are selfless, impersonal, and honest; by implication, their advice to policymakers is supposedly objective, free from personal or corporate interests, and able to withstand systematic doubt and scrutiny.

However, such an idealization of academic science, especially the view that academic scientists provide an uncontested and objective basis for policymaking, has been criticized and challenged. Among various criticisms is one made by the physicist turned humanist John Ziman. In his formulation, post-academic science produces proprietary knowledge that is focused on solving local technical problems and not necessarily made public; in the process, authority controls the activity of scientists, who are employed to achieve practical goals as experts. In the most benign form of post-academic science, the prospect of corporate funding may lure scientists away from high-risk, novel areas of research toward more readily marketable applications. Thus, post-academic science may have more subtle effects not only on the integrity of the research system but also on relevant policymaking.[25] That is, in contributing their expertise to policymaking, post-academic scientists likely bring their personal interests to bear, although doing so does not necessarily preclude them from taking societal interests into consideration.

Other scholars have also described a shift of knowledge production from individual-investigator-initiated and discipline-based "Mode 1" academic science to "Mode 2" post-academic science, in which the rise of application-focused and interdisciplinary science not only commands a set of norms for scientists different from the Mertonian ones but also has implications for policymaking.[26] As it has become increasingly difficult to establish a clear demarcation and differentiation between science and society, policymaking of post-academic science needs to involve more stakeholders or more forces—social, economic, and political—that act on science as well as on policymaking underlying or relevant to the development of science.[27]

Within the general framework of a Mode-2 society associated with Mode-2 post-academic science, there is strong pressure for policymakers to move from a more "segregated" interaction solely with the scientific community to one in which the policymaking process becomes more "integrated" with the social context.[28]

In advising on the formation of science-related policy, post-academic scientists are no longer disinterested in matters other than advancing knowledge. Instead, they become scientific arbiters, issue advocates, and honest brokers, with their exact role in policymaking often being somewhere between these, depending on the level of their interaction with society and policymakers.[29] Moreover, in a context in which the legitimacy and moral authority of scientific "facts" are subject to challenge, there is widespread recognition of the need for new mechanisms of public understanding of science, as well as public consultation and participation in policymaking, for increased accountability of science as an institution.

Meanwhile, in the early 1990s, Silvio O. Funtowicz and Jerome R. Ravetz, two scholars of environmental and technological risk and related policy, developed the concept of post-normal science. With its origin in the philosopher of science Thomas Kuhn's "normal science,"[30] post-normal science initially attempted to characterize a methodology of inquiry beyond "puzzle-solving." According to Funtowicz and Ravetz, "scientists now tackle problems introduced through policy issues where, typically, facts are uncertain, values in dispute, stakes high and decisions urgent."[31] Believing that the mistakes made by scientists in policymaking can be costly or lethal, they advocated for an "extended peer community" consisting of all the stakeholders who are affected by or have interests in a policy and are prepared to enter a dialogue on it. Stakeholders other than scientists can bring their "extended" facts, including local knowledge and other information necessary for policymakers, to assure the quality of the policymaking process.

On the other hand, as modern science and technology become increasingly powerful and ubiquitous, achieving an enhanced understanding of them is complemented and supplemented by an understanding of the possible unanticipated and undesirable consequences, which, accompanying progress, could become salient more rapidly than can be managed by existing institutions. As the knowledge society "proliferates risks—environmental, ethical and intellectual," it is necessary but very difficult, if not impossible, for different perceptions of such risks to be reconciled.[32] That is, the context of uncertainties and some long-term issues for which information is incomplete and asymmetric require involving and engaging more stakeholders, not just scientists, in the policymaking process, while also keeping the stakeholders scientifically informed and updated. This calls for new forms of governance of science for the protection of society, of the environment, and of science itself.

Policymaking pertaining to the research and especially commercialization of GM crops represents an example of policymaking involving post-academic science and post-normal science. Governments around the world have been faced with the challenge of negotiating and developing consensus around the technology that attends to the interests and concerns of various stakeholders—not only scientists, biotechnology companies, and farmers but also regulators, NGOs, and the public. As the management communication expert Alison Henderson and her colleagues have demonstrated, the GMO debates have become a public "battleground" where different stakeholders strategically compete to set the contours and agendas of the debate, to command the moral high ground, and to influence policymaking.[33]

In all GM nations, the GMO policymaking initially functioned in the interest of the academic community or was largely the purview of a policy network confined to stakeholders who generally embraced a paradigm of scientific progress.[34] That is, the policy network was established and occupied by academic scientists sharing a homogeneous perspective anchored in the progressiveness of transgenic technology and GM crops. The scientists were also closely interconnected with government bureaucracies, especially the science and agriculture ministries, as well as with MNCs and leading institutions of learning, from which they acquired new knowledge and sought collaboration. Some of them have become entrepreneurs themselves. It was only when the public was mobilized to contest the scientific-progress paradigm and policy that the governments began to rethink the appropriateness of delegating policymaking on GMOs entirely to scientists and bureaucrats.

However, as discussed in this book, in China, there was a lack of transparency in the disclosure of the composition of the National Biosafety Committee of Agricultural GMOs, as well as in the process through which MOA announced the issuance of biosafety certificates to GM rice and phytase maize, and stakeholders other than transgenic scientists have largely been excluded from policymaking. While it may be an exaggeration to say that

GM scientists have overwhelmed the national biosafety committee, as discussed in chapter 4, the simple fact is that not a single layperson or "public interest" representative has served on the committee. Although such laypersons most likely are unable to "offer any more reliable or complete a reflection of perspectives in wider society than would those of the individual experts currently involved," their mere presence would symbolize representation of "the full diversity of interest and values which characterize the different interested and affected parties."[35]

In China, the public's direct involvement in making science policy, and indeed in making any public policy, has been largely nonexistent, which explains why paranoia has been fostered and public trust in government has been undermined. The government has been perceived as not taking the public interest into consideration when making technology-related policy. Therefore, it is not surprising that there have been both rising concerns among Chinese consumers regarding their "right to know" and the public's severe suspicion toward GM foods.

However, the problematic or even seriously flawed aspects of the policy-making process have been amplified and distorted. In fact, the GMO activists have assigned distinctive political and ideological connotations to the debates and policy. In particular, they have speculated and conjectured dark and malicious motivation among the key figures—scientists and government officials—involved in policymaking. They have portrayed transgenic technology as a huge trap set up jointly by the U.S. government and Monsanto. According to the activists, the United States has provided scholarships and other opportunities to lure Chinese students and scholars for the sake of cultivating a contingent of faithful spokespersons of U.S. interests in China's biotechnology sector. The United States has also bribed key officials in relevant Chinese government ministries, hoping that they would later formulate policies in America's favor or on its behalf. Now, its dual efforts have paid off as China has commited to advancing the research and commercialization of GM rice and other crops in which American technologies are to be utilized. The activists have condemned leading Chinese agricultural biotechnologists as betraying their nation for personal gain, as well as catering to U.S. interests in conspiring to promote the commercialization of GM crops in China. But the truth is that by putting honor and reputation above personal interests, Chinese GMO scientists are also worried about the issue of biodiversity and do not want to be condemned by later generations for their irresponsibility.[36]

TABLE 8.1

R&D Expenditure for the Mega-Engineering Program of GMO New
Varieties (RMB million)

Year	2009	2010	2011	2012	2013	2014
R&D expenditure	1,090	2,920	420	830	540	665

Source: Annual final accounts of the Ministry of Agriculture.

In a word, the debates on GM crops and foods have become increasingly intense and contentious, migrating from science, its original domain, to one with societal interests, and becoming politicized. The GMO activists have succeeded to some extent in influencing, if not shaping, public opinion, which in turn has affected policy, which used to be largely promotional. The fact that biosafety certificates issued to the two strains of GM rice and phytase maize were allowed to expire is especially significant.

The seeming retreat of the policy is also reflected in an obvious decrease of funding earmarked to the MEP for GMO new varieties. Between 2009 and 2014, the program received a total of RMB6.465 billion ($982 million) from the central government; 2013 saw an investment of only RMB540 million ($87 million), an 80 percent drop from RMB2.9 billion ($431 million) in 2010. Therefore, unless the government catches up in the remaining years of the MLP period by significantly increasing appropriations, China will not meet its goal of investing some RMB20 billion in the program (table 8.1).

S&T Policy Cultures and Their Implications for China's GMO Policy

In the contemporary Chinese context, the discussion of policymaking of post-academic and post-normal science is closely related to a discussion of the competing types of policy culture—bureaucratic, economic, academic, and civic—or constellations of values, interests, and institutional preferences in the formulation of an S&T-related policy.[37] The dominant bureaucratic policy culture emphasizes the role of the state in knowledge production and application as well as the role of science and technology in nation building and national sovereignty. The bureaucratic policy culture entails a willingness of the state to use its power to prioritize and mobilize resources to the technological areas considered to be the most promising and critical.

The economic policy culture, a second type, sees the market playing an increasingly important role in technological development as well as in scientific research. The commercial value of research emphasized by this culture does not necessarily give much attention to, nor operate at the expense of, the social relevance and effects of the technology, the critical aspect of the civic policy culture. The third, or academic, policy culture defines the goal of research as the advance of knowledge for its own sake, with professional autonomy and self-governance superseding the influence of both the state and the market. Meanwhile, there is also the recognition that effective institutions of peer review require an academic tradition that cannot be established by simple policy decisions. Apparently, the academic policy culture emphasizes the role of academic scientists in policymaking.

Finally, the civic policy culture objects to leaving questions as to the uses of new technologies, especially the acceptability of risk of new technologies and the relationship between the ethical application of new technologies and enduring social values, to the state, the market, or the scientific community. Instead, it calls for expanded public participation in S&T policymaking and public debates in the assessment of such technologies. In this context, the civic policy culture aligns well with the policymaking of post-academic science and post-normal science.

The four S&T policy cultures pertain to the discussion of China's policy on GMOs. The advance of transgenic technology in China signals an enlargement of the bureaucratic policy culture's territory, reflected in the state's investment of public money in research on GM crops and in attempts to delegate the power of formulating relevant policy mainly to members of the academic community. Consequently, agricultural biotechnologists have been the main beneficiaries, not only receiving public funding for their research so as to be able to advance their profession and career but also gaining financially through commercialization of their research. Bureaucrats also have vested interests in seeing a return on the public investment to justify their initial decision to support research on agricultural biotechnology. Better linkage of research and the economy has been the mandate of the reform of the S&T system, the core of the economic policy culture. Finally, advocating the importance of the academic policy culture, biotechnologists have insisted on confining the debates on GMOs within academic science and resolving the controversies around GMOs by observing norms of academic science and normal science such as peer review and professional autonomy.

However, amid rising global concerns over the uncertainty associated with transgenic technology and GM crops, the Chinese government has also faced the challenges of balancing the promotion of research with the minimization of possible adverse effects of the relatively new technology. The issue becomes especially salient when GM crops are commercialized and GM ingredients are increasingly incorporated into foods. The rampant food safety scandals of the past decade—from melamine-tainted infant formula to disgusting gutter oil (*digouyou*) to toxic capsules for medication—have not only raised the awareness of food safety among the public, but also complicated the call of academic scientists and government's decision to commercialize more GM crops. Against this backdrop, China has established and gradually improved a biosafety regulatory regime to show that the government is on top of risk assessment and management. Doing so may overcome the ignorance of the social relevance and effects of the research on transgenic technology, the core characteristics of the bureaucratic policy culture.

Then, there has been the increasing tension between the academic community and the public and NGOs that advocate for the interests of environmental protection and biosafety. How novel technologies such as transgenic technology are socially embedded and how GMOs come to be accepted in a wider social context have posed significant challenges to science-society relations. Thus, it is necessary to involve the public in the formulation and implementation of relevant policy. This is the area where the civic S&T policy culture is supposed to come into play. By and large, the Chinese government remains ambivalent toward NGOs—viewing them, on one hand, as troublemakers because their activism seeks to arouse public concern for rights that could undermine its political control and, on the other, as helping hands mobilizing resources from society to solve problems that it is unable to solve.[38] Indeed, it is because of the campaign efforts of Greenpeace that ordinary Chinese first heard about, and then show interest in, "transgenesis" (*zhuanjiyin*). Greenpeace has been especially skillful in utilizing the media as a tool to first frame GMOs as an issue of biosafety and then reframe it as an issue of food safety to effectively attract public attention and affect public opinion.[39] Along with environmental NGOs, journalists, scholars of humanities and social sciences, and consumers were also instrumental in influencing the Chinese policymaking process regarding the development of hydropower.[40] Similarly, GMO activists have achieved their fundamental goal of halting action to approve and then proceed with the commercialization of GM rice and phytase maize. In a word, policy on GMOs has been formed and changed through "elaborate negotiations, mediations, consultations, and contestations which take place within the public arena."[41] The state has become a transgressive institution, with its bureaucratic policy culture penetrated by and penetrating academic, economic, and civic cultures of the S&T policy, as well as penetrated by and penetrating the scientific community, market, social movements, and individual concerns of consumers and citizens.

Finally, Chinese policy development regarding transgenic technology is not isolated from global trends. Suffice it to say that China's establishment of the biosafety regulatory regime was influenced by the rising global concerns over GMOs. In addition, as noted, Greenpeace has been visible in mobilizing the Chinese public against the growing of GM crops and the introduction of GM foods. Amid an overall rising Chinese nationalism in recent years have been groundless accusations against GM scientists simply because of their professional connection with U.S.-based organizations. Indeed, GMO activism has intentionally operated not only on anti-technology, anti-MNCs, and anti-globalization sentiment but also on anti-American populism. This goes beyond the discussion of post-academic science and post-normal science and S&T policy culture.

China's Policy, China's Dilemma

China's policymaking on research and commercialization of GMOs provides an example of a science policymaking process in which multiple actors/ stakeholders with multiple interests voice multiple inputs leading to discordant outcomes. The competition among these intertwined and multifaceted interests has complicated policymaking. Over time, GM crops and foods have been framed differently—as an innovative breakthrough toward progress, as a food-safety issue, as an environmental hazard, as a trade issue, or as a food-security concern.[42] These different frames have mobilized and motivated different actors/stakeholders, sometimes in a polarized and politicized way.[43] GMO activism's strategy of framing risk assessment of GM crops in China as "secret operations" and "manipulation," using the anti-American, anti-globalization, anti-MNCs, and anti-technology rhetoric, has dragged the debates away from the original spheres of biosafety and food safety with varying degrees of success. Responding to the competing frameworks and interests of various stakeholders, the Chinese government has adopted different permutations of a balanced policy by supporting research and applications in order to stay, or become, economically competitive while using regulations as means of containing possible risks.[44] However, both the pro- and anti-GMO camps have been unhappy with such a position.

On one hand, China has established a fairly rigorous biosafety regulatory regime and applied it to the risk assessment of the entire process from research to commercialization of GM crops and foods. Consequently, it took ten years for the Bt-rice varieties to undergo the entire biosafety-certification process. Although the way by which MOA disclosed the Bt-rice biosafety certification was undesirable and even questionable, China has been fairly open and transparent in disclosing information on the assessment criteria in accordance with international practice. Nonetheless, amid anti-GMO activism, the Chinese government lacked the courage to put Bt rice forward for commercialization, leading to complaints from the agricultural biotechnology research community that the prospect for commercializing a new GMO in China becomes more remote. The delay in allowing commercial cultivation of Bt rice and phytase maize and in granting biosafety certificates to other GM crops has further subjected research on GMOs to institutional uncertainty.[45] Chinese GMO scientists have complained that the results and even the existence of research on GMO biosafety have been ignored in the debates. If the government hesitated to commercialize Bt rice or phytase maize, some scientists argued, why should it have issued biosafety certificates or even invested in research at the first place? These scientists have been disappointed that the government has repeatedly called for the strengthening of linkages between research and the economy, but their efforts in commercializing research results have not been appreciated.

On the other hand, the anti-GMO camp has also been dissatisfied. The activists contend that the global debates on the risks and benefits associated with GM crops and foods remain unsettled and that decisions on the commercialization of Bt rice, a staple crop, should go beyond science. In their view, the government should have involved members of society in debate and discussion, rather than simply delegating policymaking power to scientists who have vested interests in the research, and sometimes the commercialization, of GM crops.

Finally, the government apparatus in charge of GMO policy has had its own frustrations. Entrusted by the central government to deal with the matter, the agricultural ministry has strictly followed the risk-assessment criteria and procedures outlined in the State Council regulations. It also has to be accountable for the tremendous amount of public money that has been invested in research on GMOs. However, its policymaking power has been not only challenged by both the research community and the GMO skeptics, but also constrained by the central government. In reality, the ministry seems not to possess any real authority as high-level leadership can veto its decisions. By way of comparison, in the United States, the deregulation of a GM crop is based on science and economic competitiveness. In Europe, the introduction of a preventive biosafety regulatory regime to restrict the development of GM crops and foods is mainly because of a series of food-safety scandals; those opposing GM foods have also come from the sector of organic and natural farming. In China, however, the obstacles to the development of transgenic technology and GM crops are much more mixed.

The irony is that China, like the European Union, bans the domestic cultivation of GM soybeans and corn but does allow imports of the GM varieties in large quantities for consumption. This paradoxical policy has stifled the possible development of domestic GM soybeans and corn industries to compete with and block imported GM soybeans and corn. In this regard, China seems not to have learned the lesson from its development of Bt cotton. In the 1990s, when China's cotton industry suffered severe damage from bollworm, there was an acute need for the commercialization of Bt cotton. Because of their bollworm-resistant effectiveness, Monsanto's Bt-cotton varieties soon occupied more than 80 percent of China's cotton acreage. Meanwhile, with government's support, Chinese scientists rose to the occasion with indigenously developed Bt-cotton varieties, which in less than ten years won out over Monsanto's seeds. This could be attributed to China's innovation policy favoring indigenous Bt-cotton seeds, which also have cost advantages, as well as the fact that Monsanto encountered barriers in getting its varieties approved for commercialization in China and became reluctant to introduce newer and more sophisticated seeds into China under an investment environment not in its favor.[46] By contrast, without domestically developed Bt-cotton varieties, India has conceded its Bt-cotton market to Monsanto.

However, precisely because the Chinese government has not approved the commercial cultivation of GM soybeans, China does not have domestically developed GM-soybean varieties to compete with imports so as to reduce their quantity and scale. China is said to have abandoned its efforts on developing GM soybeans.[47] Consequently, China is now the world's largest soybean importer—95 million metric tons in 2017, mostly GM varieties, seven times more than domestic non-GM soybean production.[48] The fear is that China is repeating this mistake in maize/corn. With its increasing demand for animal feeds, China since 2010 has become a net importer of corn, mostly from the United States. China also saw the expiration of the biosafety certificate issued to phytase maize in August 2014, although the certificate was renewed in early 2015 for another five years. If the commercial planting of phytase maize is further delayed, in a few years, imported GM corn will likely dominate China's corn market. This scenario also means that China will gradually lose its advantage as an early mover, as its gap with the United States and other GM nations in terms of levels of technology and varieties of GM crops on the market and in the pipeline has been enlarging rather than narrowing. The domination of Chinese dining tables by imported GM crops and by foreign companies is not what Chinese GM scientists and political leaders want to see. This likely prompted Xi Jinping, the general secretary of the Chinese Communist Party's Central Committee and China's president, to express his concern about the domination of the Chinese agricultural GM product market by foreign companies in his speech at the 2014 central conference on rural work, as discussed in chapter 3.

Not approving GM soybeans and corn for cultivation in China also deprives Chinese farmers of the benefits from growing such crops. By comparison, Argentina exports almost 100 percent of its GM soybeans, with the bulk of the benefits—about three-quarters of the $72.65 billion estimated to have been earned between 1996/97 and 2010/11—going to farmers, ten times the benefits that the agricultural industry received; its national government got the rest by way of export duties.[49]

Proceeding with research and commercialization of GMOs is closely dependent on the design and implementation of the regulatory framework governing such organisms. Transgenic technology can contribute to human well-being only if the regulatory framework is transparent, predictable, balanced, and, most important, science-based. Not only do the risks that may accompany the technology need to be addressed, but the benefits also need to be taken into account in ensuring good governance. To achieve this, policymakers, regulators, and the public need to be better informed of the objectives of and progress in transgenic technology. Conversely, the scientific community also needs to be informed of and involved in policymaking related to the technology and its implications for research so as to be better aligned with broader policy developments relating to food safety, biosafety, environmental protection, and sustainable development. Finally, under circumstances in which definitive answers do not and cannot exist in the face of uncertainties and ignorance, scientists need to carefully design studies to generate new knowledge and insights regarding GM foods to facilitate policymaking.[50]

While China's policy pertaining to transgenic technology has been evolving alongside the global development of and debates surrounding the technology, in recent years China has witnessed the introduction of conflicting domestic regulations at different levels. For example, the draft Food Law stipulates that transgenic technology should not be applied to staple crops without approval. In late 2013, the municipal government of Zhangye, Gansu Province, issued a document, "Opinions on Constructing a Safe Agricultural Product City," promoting organic farming over the growing of GM crops.[51] And Heilongjiang Province's Food Safety Regulations, which took effect on May 1, 2017, forbid not only the illegal production and operation of GM food crops and supply of GM food crops to growers and the illegal production, processing, and sale of agricultural GM products used for foods, but also the planting, production, processing, and sale of GM maize, rice, soybeans, and other crops in the province.[52]

In order to depoliticize the debates on the use of transgenic technology in agriculture and food production and make sound policy on research and commercialization of GM crops, China needs to improve communication on GM technology and crops to support science-based and participatory policymaking. However, in the marketplace of ideas, the positivist "rationalist" perspective that is usually provided by scientists and other experts is insufficient because this perspective, the so-called knowledge-deficiency model, assumes that the public is at the receiving end of new knowledge. Instead, the public needs a subjectivist view to make informed decisions in light of risk and uncertainty. That is, more comprehensive knowledge is the foundation for understanding the complicated and complex issues related to GMOs.[53]

Most recently, genome-editing technology, especially that involving CRISPR/CAS9, is revolutionizing the life sciences, including agriculture, to generate gene-edited organisms (GEOs). China is a frontrunner in this regard, with Gao Caixia, at the Chinese Academy of Sciences Institute of

Genetics and Developmental Biology, being among the first to use the technology in crops, specifically wheat and rice. After receiving a PhD from China Agricultural University in 1997, Gao went to DLF, a Danish grass seed company, first as a postdoctoral researcher and then as a research scientist. But European suspicion of GM crops "left her with little hope that her work would leave the lab." That issue, plus a desire to return with her children to her mother language and culture, sent her back to Beijing in 2009. Gao, best known for tackling genetic engineering in wheat, a crop that is legendary for its difficulty to work with, in part because many strains have six copies of the genome, was named one of the "Science Stars of China" by *Nature* in 2016.[54] She is now working with a U.S. company to further develop a disease-resistant wheat engineered at her lab. Gao also sits on the fifth National Biosafety Committee on Agricultural GMOs. With the CRISPR/CAS9 technology licensed from the Broad Institute, a partnership including Harvard University and MIT that holds a patent on the technology, Syngenta, now part of ChemChina, is developing gene-edited crops aiming not only for China but also for the global market.[55]

Ultimately, whether GM crops or gene-edited crops will find their way to the Chinese market and GM or gene-edited rice will end up on Chinese dining tables depends on whether the Chinese people are willing to accept and consume them.[56] All stakeholders—policymakers, scientists, farmers, companies, NGOs, and consumers—need to work together to come up with a consistent and sustainable policy and decide China's development of GM or gene-edited crops and foods, although the trajectory still seems tortuous and rough.

Notes

Introduction

1. RMB is expressed in current dollars (that is, the dollar value at the time) throughout the book based on average exchange of the year. See National Bureau of Statistics of China, *China Statistical Yearbook* (Beijing: China Statistics Press, various years).
2. Fang Zhouzi is the pen name of Fang Shiming, who received a Ph.D. in biochemistry from Michigan State University in 1995. Instead of launching into a scientific career as most of his overseas Chinese peers have done, Fang has engaged in a campaign against pseudoscience and fraud in science in China through New Threads, a popular website, and social-media platforms. Through writings and speeches, Fang has become an enthusiastic promoter of research on and commercialization of GM crops.
3. Cui Yongyuan, "Cui Yongyuan's Proposals: Pursuing GMOs" (in Chinese), accessed December 20, 2015, http://news.sina.com.cn/o/2014-03-03/053029605547 .shtml.
4. Clive James, *Global Status of Commercialized Biotech/GM Crops* (ISAAA Annual Briefs, 1996–2015), accessed June 1, 2016, http://www.isaaa.org/resources /publications/briefs/default.asp.
5. Ma Aiping, "More than 90 Million Tons of Soybeans Imported Last Year to Ensure Chinese People Eat Well" (in Chinese), *Science and Technology Daily*, January 31, 2018.
6. Robert L. Paarlberg, *The Politics of Precaution: Genetically Modified Crops in Developing Countries* (Baltimore: Johns Hopkins University Press, 2001).

7. Paarlberg, *The Politics of Precaution*, 93–120.

8. Daniel Lee Kleinman, Abby J. Kinchy, and Robyn Autry, "Local Variation or Global Convergence in Agricultural Biotechnology Policy? A Comparative Analysis," *Science and Public Policy* 35, no. 5 (2009): 361–371.

9. Nancy N. Chen, "Feeding the Nation: Chinese Biotechnology and Genetically Modified Food," in *Asian Biotech: Ethics and Communities of Fate*, ed. Aihwa Ong and Nancy N. Chen (Durham, NC: Duke University Press, 2010), 84.

10. Yunhe Li et al., "Biosafety Management and Commercial Use of Genetically Modified Crops in China," *Plant Cell Report* 33, no. 4 (April 2014): 570.

11. For works along such lines, see, for example, Sheila Jasanoff, *Designs on Nature: Science and Democracy in Europe and the United States* (Princeton, NJ: Princeton University Press, 2005); and Glenn Davis Stone, "The Anthropology of Genetically Modified Crops," *Annual Review of Anthropology* 39 (2010): 381–400.

1. Transgenic Technology and GMO Controversies

1. Gordon Conway and Gary Toenniessen, "Feeding the World in the Twenty-First Century," *Nature* 402 Supplement (December 2, 1999): C56.

2. Skipping some technical details of this chapter will not affect one's reading of the rest of the book.

3. Robin Marantz Henig, *A Monk and Two Peas: The Story of Gregor Mendel and the Discovery of Genetics* (London: Phoenix, 2001).

4. Nobel Foundation, "The Nobel Prize in Physiology or Medicine 1962," accessed July 28, 2014, http://www.nobelprize.org/nobel_prizes/medicine/laureates/1962/.

5. James D. Watson, *The Double Helix: A Personal Account of the Discovery of the Structure of DNA* (New York: Atheneum, 1968).

6. Paul Berg and Janet E. Mertz, "Personal Reflections on the Origins and Emergence of Recombinant DNA Technology," *Genetics* 184, no. 1 (January 2010): 9–17.

7. Steven W. Collins, *The Race to Commercialize Biotechnology: Molecules, Markets and the State in the United States and Japan* (London: Routledge, 2004), 99.

8. Collins, *The Race to Commercialize Biotechnology*, 99.

9. Berg and Mertz, "Personal Reflections"; see also Paul Berg, "Asilomar and Recombinant DNA" (Stockholm: Nobel Foundation, August 26, 2004), accessed July 29, 2014, http://www.nobelprize.org/nobel_prizes/chemistry/laureates/1980/berg-article.html. The other half of the Nobel Prize for that year went to Walter Gilbert and Frederick Sanger "for their contributions concerning the determination of base sequences in nucleic acids." For a

discussion of the episode of the discovery of rDNA technology and the associated disputes over the priority, patenting, and the award of the Nobel Prize, see Mark Jones, "The Invention of the Recombinant DNA Technology," *LSF Magazine*, Summer 2013, https://medium.com/lsf-magazine/the-invention-of-recombinant-dna-technology-e040a8a1fa22.

10. Berg and Mertz, "Personal Reflections," 13, 15–16. Beginning with the Asilomar conferences, scientists and their institutions also began to strategically control media access to information. For a discussion on that, see Matthew C. Nisbet and Bruce V. Lewenstein, "Biotechnology and the American Media: The Policy Process and the Elite Press, 1970 to 1999," *Science Communication* 23, no. 4 (June 2002): 359–391. Hepeng Jia called my attention to this point, which is much appreciated.

11. Freeman J. Dyson, *The Sun, the Genome, and the Internet: Tools of Scientific Revolution* (Oxford: Oxford University Press, 1999), 70–71.

12. Ann M. Showalter et al., "Development, Agronomic Performance and Sustainability of Transgenic Cotton for Insect Control," in *Biotechnology and Agricultural Development: Transgenic Cotton, Rural Institutions and Resource-Poor Farmers*, ed. Robert Tripp (London: Routledge, 2009), 49–51.

13. Graham Brookes, "The Economic and Environmental Impact of First-Generation Biotech Crops," in *Successful Agricultural Innovation in Emerging Economics: New Genetic Technologies for Global Food Production*, ed. David J. Bennett and Richard C. Jennings (Cambridge: Cambridge University Press, 2013), 61–81; David Rotman, "Why We Will Need Genetically Modified Foods," *MIT Technology Review* 117, no. 1 (January/February 2014): 28–37.

14. Donald H. Dean, "Biochemical Genetics of the Bacterial Insect-Control Agent *Bacillus thuringiensis*: Basic Principles and Prospects for Genetic Engineering," *Biotechnology & Genetic Engineering Reviews* 2, no. 1 (1984): 341–363.

15. Joel S. Griffitts et al., "Glycolipids as Receptors for *Bacillus thuringiensis* Crystal Toxin," *Science* 307 (February 11, 2005): 922–925.

16. David A. Victor, "Trade, Science, and Genetically Modified Foods" (New York: Council on Foreign Relations, March 13, 2001), accessed September 1, 2014, http://www.cfr.org/agricultural-policy/trade-science-genetically-modified-foods/p8689; Daniel Cressey, "A New Breed," *Nature* 497 (May 2, 2013): 27–29.

  However, according to a Chinese agricultural biotechnologist, improved quality such as taste and nutrition for GMOs belong to the second-generation while functional and therapeutic crops and foods fall into the third-generation GMOs; interview, Beijing, December 12, 2003.

17. Robert E. Black et al., "Maternal and Child Undernutrition: Global and Regional Exposures and Health Consequences," *Lancet* 371, no. 9608 (January 19, 2008): 253.

18. Frédéric Varone and Nathalie Shiffino, "Conflict and Consensus in Belgian Biopolicies: GMO Controversy Versus Biomedical Self-Regulation," in *The Politics of Biotechnology in North America and Europe: Policy Networks, Institutions, and Internationalization*, ed. Éric Montpetit, Christine Rothmayr, and Frédéric Varone (Lanham, MD: Lexington Books, 2007), 198.

19. Carl E. Pray, "Public and Private Collaboration on Plant Biotechnology in China," *AgBioForum* 2, no. 1 (1999): 49.

20. Andy Fell, "Nothing Ventured, Nothing Gained," *UC Davis Magazine* 21, no. 3 (Spring 2004), accessed June 6, 2014, http://ucdavismagazine.ucdavis.edu /issues/sp04/feature_1.html.

21. Michael Winerip, "You Call It a Tomato?," *New York Times*, June 24, 2013.

22. Clive James, *ISAAA Brief 52: Global Status of Commercialized Biotech/GM Crops: 2016*, Executive Summary, accessed June 10, 2017, http://www.isaaa .org/resources/publications/briefs/52/executivesummary/default.asp.

23. Rotman, "Why We Will Need Genetically Modified Foods."

24. Clive James, *ISAAA Brief 51: 20th Anniversary of the Global Commercialization of Biotech Crops (1996 to 2015) and Biotech Crop Highlights in 2015*, Executive Summary, accessed June 1, 2016, http://www.isaaa.org/resources/publications /briefs/51/executivesummary/default.asp.

25. Wadgy Sawahel, "Iranian Scientists Produce Country's First GM Rice," *SciDev. Net*, February 18, 2005, accessed October 10, 2014, http://www.scidev.net /global/gm/news/iranian-scientists-produce-countrys-first-gm-rice.html.

26. Ginger Pinholster, "AAAS Board of Directors: Legally Mandating GM Food Labels Could 'Mislead and Falsely Alarm Consumers,' " American Association for the Advancement of Science (AAAS), October 25, 2012, accessed April 20, 2014, http://www.aaas.org/news/aaas-board-directors-legally-mandating-gm -food-labels-could-%E2%80%9Cmislead-and-falsely-alarm.

27. Natasha Gilbert, "A Hard Look at GM Crops," *Nature* 497 (May 2, 2013): 24–26.

28. Ulrich Beck, *Risk Society: Towards a New Modernity*, trans. Mark Ritter (Los Angeles: Sage, 1992); Helga Nowotny, Peter Scott, and Michael Gibbons, *Re-Thinking Science: Knowledge and the Public in an Age of Uncertainty* (Cambridge: Polity Press, 2001), 33–34.

29. Royal Society Expert Reviewers of Pusztai Rat Experiments at the Rowett Institute from 1998–1999, "Expert Reviewers 1–6 Sent with Mail to A. Pusztai on May 10, 1999 and Revised Expert Review Sent to A. Pusztai on May 12, 1999," accessed November 26, 2014, http://www.ask-force.org/web/Pusztai /Royal-Society-Experts-1-7-19990510.pdf; Stanley Ewen and Árpád Pusztai, "Effect of Diets Containing Genetically Modified Potatoes Expressing *Galanthus nivalis* lectin on Rat Small Intestine," *Lancet* 354, no. 9187 (October 16, 1999): 1353–1354.

30. John E. Losey, Linda S. Rayor, and Maureen E. Carter, "Transgenic Pollen Harms Monarch Larvae," *Nature* 399 (May 20, 1999): 214; U.S. Department of Agriculture (USDA), Agricultural Research Service, "Q&A: Bt Corn and Monarch Butterflies," accessed August 10, 2014, http://www.ars.usda.gov/is/br/btcorn/index.html#bt4.

31. Associated Press (AP), "Shell Suppliers Give $60 Million to Taco Bell," June 8, 2001, AP News Archive, accessed August 10, 2014, http://www.apnewsarchive.com/2001/Shell-Suppliers-Give-$60M-to-Taco-Bell/id-55902df5ddc5ae7a80bb61bd1b5faa0a.

32. David Quist and Ignacio H. Chapela, "Transgenic DNA Introgressed into Traditional Maize Landraces in Oaxaca, Mexico," *Nature* 414 (November 29, 2001): 541–543; A. Piñeyro-Nelson et al., "Transgenes in Mexican Maize: Molecular Evidence and Methodological Considerations for GMO Detection in Landrace Populations," *Molecular Ecology* 18, no. 4 (February 2009): 750–761; Gilbert, "A Hard Look at GM Crops."

33. Bloomberg News, "Bayer Settles with Farmers Over Modified Rice Seeds," *New York Times*, July 2, 2011.

34. Gilles-Eric Séralini, Dominique Cellier, and Joël Spiroux de Vendômois, "New Analysis of a Rat Feeding Study with a Genetically Modified Maize Reveals Signs of Hepatorenal Toxicity," *Archives of Environmental Contamination and Toxicology* 52, no. 4 (2007): 596–602; Gilles-Eric Séralini et al., "Long Term Toxicity of a Roundup Herbicide and a Roundup-Tolerant Genetically Modified Maize," *Food and Chemical Toxicology* 50, no. 11 (2012): 4221–4231.

35. Kathryn Z. Guyton et al., "Carcinogenicity of Tetrachlorvinphos, Parathion, Malathion, Diazinon, and Glyphosate," *Lancet Oncology* 16, no. 5 (2015): 490–491.

36. Daniel Cressey, "Widely Used Herbicide Linked to Cancer," *Nature*, March 24, 2015, accessed May 1, 2015, http://www.nature.com/news/widely-used-herbicide-linked-to-cancer-1.17181.

37. Food and Agriculture Organization (FAO) of the United Nations and World Health Organization (WHO), *Summary Report of Joint FAO/WHO Meeting on Pesticide Residues* (Geneva: FAO and WHO, May 16, 2016), accessed July 30, 2017, http://www.who.int/foodsafety/jmprsummary2016.pdf?ua=1.

38. Rochelle Toplensky, "European Countries Approve Weed-Killer Glyphosate for Another Five Years," *Financial Times*, November 27, 2017.

39. National Academies of Sciences, Engineering, and Medicine, *Genetically Engineered Crops: Past Experience and Future Prospects* (Washington, DC: National Academies Press, 2016).

40. Joel Achenbach, "107 Nobel Laureates Sign Letter Blasting Greenpeace Over GMOs," *Washington Post*, June 30, 2016.

41. Heidi Ledford, "CRISPR, the Disruptor," *Nature* 522 (June 4, 2015): 20–24.

42. Stephen S. Hall, "Editing the Mushroom," *Scientific American*, 314, no. 3 (March 2016), 56–63.

43. Jennifer A. Doudna and Samuel H. Sternberg, *A Crack in Creation: Gene Editing and the Unthinkable Power to Control Evolution* (Boston: Houghton Mifflin Harcourt, 2017), 119–128; Caixia Gao, "The Future of CRISPR Technologies in Agriculture," *Nature Reviews Molecular Cell Biology* 19, no. 5 (May 2018): 275–276; Maywa Montenegro, "CRISPR Is Coming to Agriculture—with Big Implications for Food, Farmers, Consumers and Nature," *Ensia*, January 28, 2016, accessed June 1, 2016, http://ensia.com/voices /crispr-is-coming-to-agriculture-with-big-implications-for-food-farmers -consumers-and-nature/.

44. Ledford, "CRISPR, the Disruptor."

2. Global GMO Policy

1. Organisation for Economic Co-operation and Development (OECD), *Recombinant DNA Safety Considerations: Safety Considerations for Industrial, Agricultural and Environmental Applications of Organisms Derived by Recombinant DNA Techniques* (Paris: OECD, 1986).

2. U.S. Office of Science and Technology Policy (OSTP), "Coordinated Framework for Regulation of Biotechnology," *Federal Register* 50 (June 26, 1986): 23302–23393.

3. Francis Garon and Éric Montpetit, "Different Paths to the Same Result: Explaining Permissive Policies in the USA," in *The Politics of Biotechnology in North America and Europe: Policy Networks, Institutions, and Internationalization*, ed. Éric Montpetit, Christine Rothmayr, and Frédéric Varone (Lanham, MD: Lexington Books, 2007), 67.

4. Dave Toke, *The Politics of GM Food: A Comparative Study of the UK, USA and EU* (Abingdon, UK: Routledge, 2004), 108–109; Garon and Montpetit, "Different Paths to the Same Result," 66.

5. OSTP, "Coordinated Framework for Regulation of Biotechnology."

6. U.S. Office of Science and Technology Policy (OSTP), "Exercise of Federal Oversight Within Scope of Statutory Authority: Planned Introductions of Biotechnology Products Into the Environment," *Federal Register* 59 (February 27, 1992): 6753.

7. "Modernizing the Regulatory System for Biotechnology Products: Final Version of the 2017 Update to the Coordinated Framework for the Regulation of Biotechnology." Accessed May 13, 2013. https://obamawhitehouse. archives.gov/sites/default/files/microsites/ostp/2017_coordinated_frame work_update.pdf, 1.

8. The discussion of the GMO regulatory framework in the United States draws extensively on David Vogel, *The Politics of Precaution: Regulating Health, Safety, and Environmental Risks in Europe and the United States* (Princeton, NJ: Princeton University Press, 2012), 73–85; Garon and Montpetit, "Different Paths to the Same Result"; and Gabriela Pechlaner, *Corporate Crops: Biotechnology, Agriculture, and the Struggle for Control* (Austin: University of Texas Press, 2012), 54–56.

9. U.S. Food and Drug Administration (FDA), "Statement of Policy: Food Derived from New Plant Varieties," *Federal Register* 57 (May 29, 1992): 22984–23305.

10. National Research Council, Committee on Genetically Modified Pest-Protected Plants, *Genetically Modified Pest-Protected Plants: Science and Regulation* (Washington, DC: National Academy Press, 2000); National Academies of Sciences, Engineering, and Medicine, *Genetically Engineered Crops: Past Experience and Future Prospects* (Washington, DC: National Academies Press, 2016).

11. U.S. Food and Drug Administration (FDA), "Premarket Notice Concerning Bioengineered Food," *Federal Register* 66, no. 12 (January 18, 2001): 4706–4738; Vogel, *The Politics of Precaution*, 73–74.

12. U.S. Environmental Protection Agency (EPA), "Regulation of Biotechnology for Use in Pest Management," accessed November 10, 2017, https://www.epa.gov/regulation-biotechnology-under-tsca-and-fifra/epas-regulation-biotechnology-use-pest-management.

13. U.S. Department of Agriculture (USDA), Animal and Plant Health Inspection Service, "Petitions for Determination of Nonregulated Status," accessed November 10, 2017, http://www.aphis.usda.gov/biotechnology/petitions_table_pending.shtml.

14. U.S. Food and Drug Administration (FDA), "Biotechnology Consultations on Food from GE Plant Varieties," accessed November 10, 2017, http://www.accessdata.fda.gov/scripts/fdcc/?set=Biocon.

15. EPA, "Regulation of Biotechnology for Use in Pest Management."

16. Mark F. Cantley and Drew L. Kershen, "Regulatory Systems and Agricultural Biotechnology," in *Successful Agricultural Innovation in Emerging Economies: New Genetic Technologies for Global Food Production*, ed. David J. Bennett and Richard C. Jennings (Cambridge: Cambridge University Press, 2013), 271.

17. See, for example, Center for Food Safety, "USDA Aims to Commercialize Genetically Engineered Sugar Beets, Ignores Documented Environmental Impacts," Washington, DC: Center for Food Safety, December 10, 2010, accessed November 10, 2017, https://www.centerforfoodsafety.org/press-releases/817/usda-aims-to-commercialize-genetically-engineered-sugar-beets-ignores-documented-environmental-impacts.

18. Cantley and Kershen, "Regulatory Systems and Agricultural Biotechnology," 271.

19. Cantley and Kershen, "Regulatory Systems and Agricultural Biotechnology," 271.

20. Garon and Montpetit, "Different Paths to the Same Result," 68.

21. Vogel, *The Politics of Precaution*, 83–84.

22. Toke, *The Politics of GM Food*, 120.

23. Vogel, *The Politics of Precaution*, 83–84.

24. Allison Kopicki, "Strong Support for Labeling Modified Foods," *New York Time*, July 23, 2013.

25. Vermont's unrestricted GMO labeling bill took effect in July 2016 as scheduled, making it the first U.S. state to require food makers to label products made from GM ingredients, such as corn, soybeans, or sugar beets. Maine and Connecticut also passed GMO-labeling laws, but both laws have "trigger" clauses providing that they will not take effect unless several neighboring states do the same. See Molly Ball, "Want to Know If Your Food Is Genetically Modified?," *Atlantic*, May 14, 2014, accessed September 10, 2014, http://www.theatlantic.com/features /archive/2014/05/want-to-know-if-your-food-is-genetically-modified/370812/.

26. Center for Food Safety. "GE Food Labeling: States Take Action." Washington, DC: Center for Food Safety, June 2014, accessed September 10, 2014. http:// www.centerforfoodsafety.org/files/ge-state-labe.ling-fact-sheet-620141_28179 .pdf.

27. Richard Levick, "Colorado and Oregon: The GMO Conversation Shifts from Risks to Benefits," *Forbes*, November 5, 2014, accessed November 20, 2014, http://www.forbes.com/sites/richardlevick/2014/11/05/colorado-and-oregon -the-gmo-conversation-shifts-from-risks-to-benefits/.

28. Carolyn Raab and Deana Grobe, "Labeling Genetically Engineered Food: The Consumer's Right to Know?," *AgBioForum* 6, no. 4 (2003): 155–161.

29. Carey Gillam, "U.S. Food Makers to Seek Single Federal Standard for GMO Labeling," *Reuters*, January 13, 2014, accessed August 10, 2014, http://www .reuters.com/article/2014/01/13/usa-gmo-labeling-idUSL2N0KN1NX20140113.

30. Center for Food Safety. "GE Food Labeling: States Take Action." Washington, DC: Center for Food Safety, June 2014. Accessed September 10, 2014. http://www.centerforfoodsafety.org/files/ge-state-labe.ling-fact-sheet-620141 _28179.pdf.

31. Gillam, "U.S. Food Makers to Seek Single Federal Standard."

32. Carey Gillam, "House Passes Anti-GMO Labeling Law," *Reuters*, July 23, 2015, accessed August 18, 2015, http://www.reuters.com/article/2015/07/23 /us-usa-gmo-labeling-idUSKCN0PX17920150723.

33. Mary Clare Jalonick, "Congress OKs Bill Requiring First GMO Food Labels," *Washington Post*, July 14, 2016.

34. Tanya Lewis, "Chipotle and GMO Debate," *Washington Post*, May 4, 2015.

35. Laura Parker, "The GMO Labeling Battle Is Heating Up—Here's Why," *National Geographic*, January 12, 2014, accessed August 10, 2014, http://news .nationalgeographic.com/news/2014/01/140111-genetically-modified-organisms -gmo-food-label-cheerios-nutrition-science/.

36. Richard Levick, "Are GMO-Free Cheerios the First Dominos?," *Forbes*, January 9, 2014; Dale Buss, "American Still Aren't Buying GMO-Free Gospel. Just Ask General Mills," *Forbes*, February 20, 2014.

37. Vogel, *The Politics of Precaution*, 63–64.

38. Helge Torgersen et al., "Promise, Problems and Proxies: Twenty-Five Years of Debate and Regulation in Europe," in *Biotechnology: The Making of a Global Controversy*, ed. Martin W. Bauer and George Gaskell (Cambridge: Cambridge University Press, 2002), 62.

39. Torgersen et al., "Promise, Problems and Proxies," 62–64.

40. Vogel, *The Politics of Precaution*, 39, 73.

41. European Commission, "GMO Legislation," accessed June 1, 2016, http:// ec.europa.eu/food/plant/gmo/legislation/index_en.htm.

42. Commission of the European Communities, *White Paper on Food Safety* (Brussels: European Union, January 12, 2000).

43. Claire Waterton and Brain Wynne, "Knowledge and Political Order in the European Environment Agency," in *States of Knowledge: The Co-Production of Science and Social Order*, ed. Sheila Jasonoff (London: Routledge, 2004), 96–98.

44. Arco Timmermans, "Accommodation, Bureaucratic Politics, and Suprana- tional Leviathan: ART and GMO Policy-Making in the Netherlands," in *The Politics of Biotechnology in North America and Europe: Policy Networks, Institutions, and Internationalization*, ed. Éric Montpetit, Christine Rothmayr, and Frédéric Varone (Lanham, MD: Lexington Books, 2007), 185, 187.

45. The following discussion draws mainly from Gemma Masip et al., "Paradoxical EU Agricultural Policies on Genetically Engineered Crops," *Trends in Plant Science* 18, no. 6 (June 2013): 312.

46. Per Pinstrup-Andersen and Ebbe Schioler, *Seeds of Contention: World Hunger and the Global Controversy Over GM Crops* (Baltimore: Johns Hopkins University Press, 2000).

47. Torgersen et al., "Promise, Problems and Proxies," 59.

48. Leo Cendrowicz, "Is Europe Finally Ready for Genetically Modified Foods?," *Time*, March 9, 2010.

49. European Commission, *A Decade of EU-Funded GMO Research (2001–2010)* (Luxembourg: Publications Office of the European Union, 2010), 262–263.

50. European Commission, *A Decade of EU-Funded GMO Research*, 240–245.

51. European Commission, *A Decade of EU-Funded GMO Research*, 241 and 209.

52. ISAAA, *Brief 52: Global Status of Commercialized Biotech/GM Crops: 2016.* Executive Summary, accessed June 10, 2017, http://www.isaaa.org/resources /publications/briefs/52/executivesummary/default.asp.

53. "EU Court Bans BASF's 'Amflora' GM Potato, Annuls Commission Approval," *DW*, December 13, 2013, accessed November 10, 2014, http:// www.dw.de/eu-court-bans-basfs-amflora-gm-potato-annuls-commission -approval/a-17293624.

54. European Commission, *A Decade of EU-Funded GMO Research*, 16.

55. European Commission, *EC-Sponsored Research on Safety of Genetically Modified Organisms: A Review of Results* (Luxembourg: Publications Office of the European Union, 2001), 12.

56. "EU Science Advisor: 'Lots of Policies Are Not Based on Evidence,'" *Euractiv*, July 24, 2012, accessed August 31, 2014, http://www.euractiv.com /innovation-enterprise/chief-scientifc-adviser-policy-p-interview-514074.

57. Guy Poppy, "Why Europe Needs a Chief Science Advisor," *Conversation*, November 17, 2014, accessed November 23, 2014, http://theconversation .com/why-europe-needs-a-chief-scientific-advisor-34313.

58. Louise O. Fresco, "The GMO Stalemate in Europe," *Science* 329 (February 22, 2013): 883.

59. Brian Heap, "Europe Should Rethink Its Stance on GM Crops," *Nature* 498 (June 27, 2013): 409.

60. Adam Vaughan, "French Ban of Monsanto GM Maize Rejected by EU," *Guardian*, May 22, 2014.

61. Barbara Casassus, "Study Linking GM Maize to Rat Tumours Is Retracted," *Nature*, November 28, 2013, accessed August 31, 2014, http://www.nature .com/news/study-linking-gm-maize-to-rat-tumours-is-retracted-1.14268.

62. Daniel Cressey, "Debate Rages Over Herbicide's Cancer Risk," *Nature*, November 15, 2015, accessed November 20, 2015, http://www.nature.com /news/debate-rages-over-herbicide-s-cancer-risk-1.18794.

63. Tania Rabesandratana, "E.U. Moves Closer to Enabling National Bans on GM Crops," *Science* (November 12, 2014), accessed November 23, 2014, http://news.sciencemag.org/environment/2014/11/e-u-moves-closer -enabling-national-bans-gm-crops.

64. Tom Mitchell, Christian Shepherd, and Ralph Atkins, "Syngenta Offers ChemChina Seeds of GM Growth," *Financial Times*, February 3, 2016; Mara Hvistendahl, "China Aims to Sow a Revolution with GM Seed Takeover," *Science* 356 (April 7, 2017): 16–17.

65. Leslie Picker, Danny Hakim, and Michael J. de la Merced, "Bayer Deal for Monsanto Follows Agribusiness Trend, Raising Worries for Farmers," *New York Times*, September 14, 2016.

66. Mark Lynas, "Lecture to Oxford Farming Conference," January 3, 2013, accessed June 1, 2016, http://www.marklynas.org/2013/01/lecture-to-oxford-farming-conference-3-january-2013/.

67. Lynas, "Lecture to Oxford Farming Conference."

68. Mark Lynas, "GM Won't Yield a Harvest for the World," *Guardian*, June 19, 2008.

69. Lynas, "Lecture to Oxford Farming Conference."

70. John Vidal and Hanna Gersmann, "Biotech Group Bids to Recruit High-Profile GM 'Ambassadors,' " *Guardian*, October 20, 2011.

71. This section draws extensively on Santosh K. Singh, *India: Agricultural Biotechnology Annual, 2011*, GAIN Report Number IN1187 (Washington, DC: USDA Foreign Agricultural Service, Global Agricultural Information Network [GAIN], September 15, 2011); *India: Agricultural Biotechnology Annual, 2012*, GAIN Report Number IN2098 (Washington, DC: USDA Foreign Agricultural Service, Global Agricultural Information Network [GAIN], July 17, 2012); *India: Agricultural Biotechnology Annual, 2013*, GAIN Report Number IN3083 (Washington, DC: USDA Foreign Agricultural Service, Global Agricultural Information Network [GAIN], July 15, 2013); *India: Agricultural Biotechnology Annual, 2014*, GAIN Report Number IN4059 (Washington, DC: USDA Foreign Agricultural Service, Global Agricultural Information Network [GAIN], July 11, 2014); *India: Agricultural Biotechnology Annual, 2015*, GAIN Report Number IN5088 (Washington, DC: USDA Foreign Agricultural Service, Global Agricultural Information Network [GAIN], July 10, 2015); and *India: Agricultural Biotechnology Annual, 2016*, GAIN Report Number IN6157 (Washington, DC: USDA Foreign Agricultural Service, Global Agricultural Information Network [GAIN], December 14, 2016).

72. Singh, *India: Agricultural Biotechnology Annual, 2015*; *India: Agricultural Biotechnology Annual, 2016*.

73. Singh, *India: Agricultural Biotechnology Annual, 2011*.

74. Chavali Kameswara Rao, "Genetically Engineered Crops Would Ensure Food Security in India," in *Successful Agricultural Innovation in Emerging Economics: New Genetic Technologies for Global Food Production*, ed. David J. Bennett and Richard C. Jennings (Cambridge: Cambridge University Press, 2013), 174.

75. ISAAA, *Global Status of Commercialized Biotech/GM Crops: 2016*.

76. Singh, *India: Agricultural Biotechnology Annual, 2016*.

77. Singh, *India: Agricultural Biotechnology Annual, 2014*.

78. Singh, *India: Agricultural Biotechnology Annual, 2014*; *India: Agricultural Biotechnology Annual, 2016*.

79. Singh, *India: Agricultural Biotechnology Annual, 2013*.

80. Singh, *India: Agricultural Biotechnology Annual, 2015*.

81. Singh, *India: Agricultural Biotechnology Annual, 2013*; *India: Agricultural Biotechnology Annual, 2015*; *India: Agricultural Biotechnology Annual, 2016*.

82. "Q&A: Manmohan Singh: India's Scholar–Prime Minister Aims for Inclusive Development," *Science* 335 (February 24, 2012): 907–908.

83. Singh, *India: Agricultural Biotechnology Annual, 2015*.

84. Singh, *India: Agricultural Biotechnology Annual, 2016*.

85. Rao, "Genetically Engineered Crops Would Ensure Food Security in India," 179.

86. Rao, "Genetically Engineered Crops Would Ensure Food Security in India," 177–178.

87. Rao, "Genetically Engineered Crops Would Ensure Food Security in India," 182.

88. Jyotika Sood, "Whose Germplasm Is It?," *Down to Earth*, October 15, 2012, accessed August 14, 2015, http://www.downtoearth.org.in/coverage/whose -germplasm-is-it-39206.

    However, Bt brinjal was approved for commercial release in neighboring Bangladesh in 2013 and is now under consideration in the Philippines.

89. Quoted in Michael Spector, "Seeds of Doubt: An Activist's Controversial Crusade Against Genetically Modified Crops," *New Yorker*, August 25, 2014.

90. Vandana Shiva, "From Seeds of Suicide to Seeds of Hope: Why Are Indian Farmers Committing Suicide and How Can We Stop This Tragedy?," *HuffPost*, accessed August 31, 2014, http://www.huffingtonpost.com/vandana-shiva /from-seeds-of-suicide-to_b_192419.html; Cy Gonick and Andrea Levy, "An Interview with Vandana Shiva," *Canadian Dimension* 48, no. 4 (July/August 2014), accessed August 31, 2014, http://canadiandimension.com/articles/6235/.

91. Rao, "Genetically Engineered Crops Would Ensure Food Security in India," 180.

92. Keith Kloor, "The GMO-Suicide Myth," *Issues in Science and Technology*, 30, no. 2 (Winter 2014), accessed October 20, 2014, http://issues.org/30–2 /keith/; see also Glenn Davis Stone, "The Anthropology of Genetically Modified Crops," *Annual Review of Anthropology* 39 (2010): 390–391.

93. Kloor, "The GMO-Suicide Myth."

94. Spector, "Seeds of Doubt."

95. European Commission, "Fact Sheet: Questions and Answers on EU's Policies on GMOs," April 25, 2015, accessed June 1, 2016, http://europa.eu/rapid /press-release_MEMO-15-4778_en.htm.

96. Frédéric Varone, Christine Rothmayr, and Éric Montpetit, "Comparing Biotechnology Policy in Europe and North America: A Theoretical Framework," in *The Politics of Biotechnology in North America and Europe: Policy Networks, Institutions, and Internationalization*, ed. Éric Montpetit, Christine Rothmayr, and Frédéric Varone (Lanham, MD: Lexington Books, 2007), 1–34.

97. Rachel Feltman, "Why This Genetically Modified Mushroom Gets to Skip USDA Oversight," *Washington Post,* April 18, 2016.

98. XiaoZhi Lim, "European Scientists Call for GMO Legislation Revamp to Prevent Stifling Genome Editing," *Genetic Literacy Project,* July 25, 2014, accessed June 1, 2016, https://www.geneticliteracyproject.org/2014/07/25 /european-scientists-call-for-gmo-legislation-revamp-to-prevent-stifling -genome-editing/.

99. USDA, "Secretary Perdue Issues USDA Statement on Plant Breeding Innovation" (Washington, DC: USDA press release, March 28, 2018), accessed April 21, 2018, https://www.usda.gov/media/press-releases/2018/03/28/secretary -perdue-issues-usda-statement-plant-breeding-innovation.

100. Matthias Fladung, "Cibus' Herbicide-Resistant Canola in European Limbo," *Nature Biotechnology* 34 (May 2016), 473–474; "Gene Editing in Legal Limbo in Europe," *Nature* 542, (February 22, 2017): 392.

3. Research and Commercialization of GM Crops in China

1. Jikun Huang et al., *Economic Impacts of Transgenic Biotechnology: Ten Years of Bt-Cotton in China* (in Chinese) (Beijing: Science Press, 2010), 21–26. See also Joseph Wong, *Betting on Biotech: Innovation and the Limits of Asia's Developmental State* (Ithaca, NY: Cornell University Press, 2011), for the "bet" metaphor used in his study of biotechnology in Taiwan, Korea, and Singapore.

2. Carl E. Pray, "Public and Private Collaboration on Plant Biotechnology in China," *AgBioForum* 2, no. 1 (1999): 49; Nancy N. Chen, "Feeding the Nation: Chinese Biotechnology and Genetically Modified Food," in *Asian Biotech: Ethics and Communities of Fate,* ed. Aihwa Ong and Nancy N. Chen (Durham, NC: Duke University Press, 2010), 83.

3. Changyong Wang, Zhidi Yu, and Dehui Wang, "China: Risk Assessment and Risk Management," in *Risk Assessment and Risk Management in Implementing the Cartagena Protocol: Proceedings of Asia Regional Workshop,* ed. Balakrishna Pisupati, Bhujangarao D. Dharmaji, and Emilie Warner (Homagama, Sri Lanka: Karunaratne & Sons, 2003), 90–101.

4. Yunhe Li et al., "Biosafety Management and Commercial Use of Genetically Modified Crops in China," *Plant Cell Report* 33 (2014): 570. However, it seems that a virus-resistant variety of tomato was also approved for commercialization in 1997. See Huang et al., *Economic Impacts of Transgenic Biotechnology,* 20.

5. Kenneth Lieberthal and Michel Oksenberg, *Policy Making in China: Leaders, Structures, and Processes* (Princeton, NJ: Princeton University Press, 1988).

6. Can Huang et al., "Organization, Program and Structure: Analysis of the Chinese Innovation Policy Framework," *R&D Management* 34, no. 4 (2004): 367–387.

7. NPC's relevance to the GM-crop law-making process includes a 2010 public petition calling for the withdrawal of biosafety certificates issued to Bt rice and a referendum to determine whether to proceed with GM-rice commercialization. See chapter 5 for further discussion.

8. Alana Boland, "The Three Gorges Debate and Scientific Decision-Making in China," *China Information* 13, no. 1 (1998): 25–42; Dai Qing, *Yangtze! Yangtze!*, ed. Patricia Adams and John Thibodeau, trans. Nancy Liu, Wu Mei, Sun Yougeng, and Zhang Xiaogang (London: Earthscan, 1994).

9. A government reorganization was introduced in the first session of the thirteenth NPC in March 2018, which affects some of the agencies mentioned here. They include the Ministry of Agriculture and Rural Affairs (formerly the Ministry of Agriculture); the National Health Commission (formerly the National Health and Family Planning Commission), and the Ministry of Ecology and Environment (formerly the Ministry of Environmental Protection). Both the former and the new names of the agencies are used in the book to reflect their relevant roles in making GMO-related policies and regulations.

10. Yutao Sun and Cong Cao, "Demystifying Central Government R&D Spending in China," *Science* 345 (August 29, 2014): 1006–1008.

11. Lieberthal and Oksenberg, *Policy Making in China*.

12. Richard P. Suttmeier and Cong Cao, "China Faces the New Industrial Revolution: Assessing Research and Innovation Strategies for the 21st Century," *Asian Perspective* 23, no. 3 (1999): 153–200; Cong Cao, "Strengthening China Through Science and Education: China's New Development Strategy Toward the Twenty-First Century," *Issues & Studies* 38, no. 3 (2002): 122–149; Cong Cao et al., "Reforming China's S&T System," *Science* 341 (August 2, 2013): 460–462; Cong Cao and Richard P. Suttmeier, "Challenges of S&T System Reform in China," *Science* 335 (March 10, 2017): 1019–1021.

13. Cao et al., "Reforming China's S&T System."

14. Feng-chao Liu et al., "China's Innovation Policies: Evolution, Institutional Structure, and Trajectory," *Research Policy* 40, no. 7 (2011): 917–931.

15. Liu et al., "China's Innovation Policies."

16. Shaoguang Wang, "Changing Models of China's Policy Agenda Setting," *Modern China* 34, no. 1 (2008): 56–87.

17. Wan Li, "Democratization and Scientification of Policy-Making Is an Important Subject of the Reform of the Political System" (in Chinese), *People's Daily*, August 15, 1986. For a discussion of Wan Li's speech, see Merle Goldman, *Sowing the Seeds of Democracy in China: Political Reform in the Deng Xiaoping Era* (Cambridge, MA: Harvard University Press, 1994).

18. Liu Li, *Theories and Practice on the Formulation of Basic Research-Related Policy* (in Chinese) (Beijing: Tsinghua University Press, 2007), 239–247.

19. The code 863 indicates that the program was initiated in March 1986.

20. Richard P. Suttmeier, Cong Cao, and Denis Fred Simon, " 'Knowledge Innovation' and the Chinese Academy of Sciences," *Science* 312 (April 7, 2006): 58–59; Richard P. Suttmeier, Cong Cao, and Denis Fred Simon, "China's Innovation Challenge and the Remaking of the Chinese Academy of Sciences," *Innovations: Technology, Governance, Globalization* 1, no. 3 (2006): 78–97.

21. Cong Cao, Richard P. Suttmeier, and Denis Fred Simon, "China's 15-Year Science and Technology Plan," *Physics Today* 59, no. 12 (2006): 38–43.

22. Cong Cao, *China's Scientific Elite* (London: RoutledgeCurzon, 2004); Cong Cao and Richard P. Suttmeier, "China's 'Brain Bank': Leadership and Elitism in Chinese Science and Engineering," *Asian Survey* 39, no. 3 (1999): 525–559.

23. Richard P. Suttmeier, "Engineers Rule, OK?," *New Scientist*, November 10, 2007, 71–73.

24. Mao Yufei, "Demystifying the Secret of Ruling Hidden in the Red-Head Official Documents" (in Chinese), *Global Personality*, August 2014, accessed August 15, 2014, http://news.sina.com.cn/c/sd/2014-08-15/120730690491.shtml.

25. Reports have given various expenditure figures for the MEP for GMO new varieties. The figure used here is based on information drawn from an interview with an agricultural economist, Beijing: November 3, 2010.

26. Chen Yu, "Experts Interpret Central Document No. 1: Push Forward Commercialization of GMO New Varieties" (in Chinese), *Science and Technology Daily*, February 3, 2010.

27. Molecular breeding, or molecular plant breeding, is defined as the use of advanced molecular biology techniques for crop improvement. It includes natural variation arising as a result of the accumulation of random mutations or polymorphisms; induced variation caused by exposure to ionizing radiations or chemical agents that increase the number of mutations in DNA; and engineered variation caused by what is commonly referred to as genetic modification (GM). See Ian Graham, "Using Molecular Breeding to Improve Orphan Crops for Emerging Economies," in *Successful Agricultural Innovation in Emerging Economies: New Genetic Technologies for Global Food Production*, ed. David J. Bennett and Richard C. Jennings (Cambridge: Cambridge University Press, 2013), 96.

28. Qiang Wang, "Chinese Scientists Must Engage the Public on GM," *Nature* 519 (March 5, 2015): 7.

29. Lü Yongyan, "Basic Research and Molecular Breeding in Central Document No. 1" (in Chinese), accessed January 23, 2014, http://tech.gmw.cn/2014-01/22/content_10196782.htm.

30. Chen Xiwen, "Interpreting the Text of Central Document No. 1" (in Chinese), January 22, 2014, accessed January 23, 2014, http://china.caixin.com/2014-01-22/100632019.html.

31. State Council Information Office, "News Conference on the Acceleration of Agricultural Modernization" (in Chinese), February 3, 2015, accessed January 30, 2016, http://www.scio.gov.cn/xwfbh/xwbfbh/wqfbh/2015/20150203/tw32541/Document/1393981/1393981.htm.

32. State Council Information Office, "News Conference to Interpret Central Document No. 1: Key Points" (in Chinese), January 28, 2016, accessed January 30, 2016, http://www.scio.gov.cn/zxbd/wz/Document/1466802/1466802.htm.

33. Xi Jinping, "China Should Occupy the High Ground Transgenic Technology" (in Chinese), September 28, 2014, accessed September 28, 2014, http://politics.caijing.com.cn/20140928/3711853.shtml.

34. Zhou Yingfeng, "Xi Jinping Participated in Popularization of Science Activities at China Agricultural University" (in Chinese), *Xinhua News Agency*, September 15, 2012, accessed September 28, 2014, http://www.gov.cn/ldhd/2012-09/15/content_2225485.htm.

35. Interview with a science journalist, Beijing, April 22, 2014.

36. Interview with a transgenic scientist, Wuhan, August 29, 2012.

37. Of the seven original fields, space and laser technologies were administered by the Commission for Science, Technology and Industry for National Defense. See Ru Peng, *Science and Technology Experts and Decision Making in Science and Technology: The Influence of Science and Technology in the Decision Making of the 863 Program* (in Chinese) (Beijing: Tsinghua University Press, 2012), 55–56.

38. Valerie J. Karplus and Xingwang Deng, *Agricultural Biotechnology in China: Origins and Prospects* (New York: Springer, 2008), 60.

39. Wu Shihong, "Deng Xiaoping's Thinking on Modernization of Agriculture in China" (in Chinese), *Literature of the Chinese Communist Party* no. 6 (November 2006): 64.

40. The 863 Program continued after the first fifteen years, and the civilian areas between 2001 and 2005 included biotechnology and modern agriculture, information, advanced manufacturing and automation, advanced energy, new materials, and resources and the environment. Under biotechnology and modern agriculture were bioengineering, gene manipulation, bioinformatics, and modern agriculture. The 863 Program during its first fifteen years also had a special subprogram that included the rice genome in agriculture. Ru, *Science and Technology Experts and Decision Making*, 210.

41. Laurence Schneider, *Biology and Revolution in Twentieth-Century China* (Lanham, MD: Rowan & Littlefield, 2003), 243–244.

42. The code 973 indicates that the program was initiated in March 1997.

43. Cao, "Strengthening China Through Science and Education."

44. Cao, Suttmeier, and Simon, "China's 15-Year Science and Technology Plan." Other MSPs include quantum research and nanotechnology, with global climate change and stem cell added later.

45. Zhang Qifa et al., "Suggestions on the Development Strategy of Research and Commercialization of GM Crops in China" (in Chinese), *Bulletin of the Chinese Academy of Sciences* 19, no. 5 (2004): 330–332; see also Liu Jianqiang, "GM Rice for the 1.3 Billion Chinese: A Game of Safety and Interests" (in Chinese), *Southern Weekend*, December 9, 2004.

46. Interview with a plant geneticist, Wuhan, August 29, 2012.

47. Mao Lei and Liao Wengen, "Wu Bangguo Chairs the NPC Standing Committee's Sixteen Seminar" (in Chinese), *People's Daily*, June 26, 2010.

48. Liang Wei, "300 Netizens Tried Golden Rice in Wuhan" (in Chinese), *Southern Metropolitan News*, October 20, 2013; Bao Xiaodong, Li Yifan, and Wang Yue, "An Unexpected Disturbance Following a Joint Petition: Looking for 61 'Pro-GMO' Academicians" (in Chinese), *Southern Weekend*, December 5, 2013.

49. Hao Xin and Richard Stone, "Q&A: China's Science Premier," *Science* 322 (October 17, 2008): 363.

50. Wen Jiabao, "Report on the Work of the Government," March 5, 2010, accessed June 20, 2014, http://www.npc.gov.cn/englishnpc/Speeches/2010-03/19/content_1564308.htm.

51. Rao Yi, "Development and Challenges of Chinese Science: Publications in International Life Sciences Journals" (in Chinese), *Twenty-First Century* no. 1 (February 2002): 83–93; Schneider, *Biology and Revolution in Twentieth-Century China.*

52. Cong Cao, "Chinese Science and the 'Nobel Prize Complex,' " *Minerva* 42, no. 2 (2004): 151–172.

53. Rao Yi, Zhang Daqing, and Li Runhong, *Tu Youyou and Artemisinin* (in Chinese) (Beijing: China Science and Technology Publisher, 2015).

54. Jun Yu, "Biotechnology Research in China—A Personal Perspective," in *Innovation with Chinese Characteristics: High-Tech Research in China*, ed. Linda Jakobson (New York: Palgrave McMillan, 2007), 134–165.

55. James Keeley, "The Biotech Developmental State? Investigating the Chinese Gene Revolution," IDS Working Paper 207 (Brighton, UK: University of Sussex, Institute of Development Studies, 2003), 13.

56. I would like to thank Yutao Sun for locating this information.

57. Barbara Nasto, "Enter the Dragon: China's Biopharmaceutical Clusters," *Biotech in China, Nature Biotechnology* Supplement (December 2012): B5.

58. Schneider, *Biology and Revolution in Twentieth-Century China*, 78–85; Cao, *China's Scientific Elite*, 91.

59. See, for example, Sharon LaFraniere, "Fighting Trend, China Is Luring Scientists Home," *New York Times*, January 6, 2012.

60. Yu, "Biotechnology Research in China."

61. Ping Zhou and Wolfgang Glanzel, "In-depth Analysis on China's International Collaboration in Science," *Scientometrics* 82, (2010): 597–612.

62. Nan Ma and Jiancheng Guan, "An Exploratory Study on Collaboration Profiles of Chinese Publications in Molecular Biology," *Scientometrics* 65, no. 3 (2005): 343–355.

63. Zhu Chen et al., "Life Sciences and Biotechnology in China," *Philosophical Transactions of the Royal Society, B: Biological Sciences* 362 (2007): 947–957.

64. Wang Danhong, "Shi Yigong: The Twenty-First Century Is the Century of the Life Sciences" (in Chinese), *Science Times*, August 7, 2007, A3.

65. Interview, Wuhan, August 29, 2012.

66. Interview, Beijing, November 3, 2010.

67. Mao Chen, Anthony Shelton, and Gong-yin Ye, "Insect-Resistant Genetically Modified Rice in China: From Research to Commercialization," *Annual Review of Entomology* 56 (2011): 86.

68. Qifa Zhang, "China," in *Agricultural Biotechnology: Country Case—A Decade of Development*, ed. Gabrielle J. Persley and L. Reginald MacIntyre (Wallingford, UK: CABI, 2001), 41–50.

69. Jikun Huang et al., "Plant Biotechnology in China," *Science* 295 (January 25, 2002): 674–677.

70. Ministry of Science and Technology, "China's Plant GMO Research and Commercialization Special Program Achieve Significant Progress" (in Chinese), April 16, 2007, accessed June 21, 2014, http://www.most.gov.cn/ncs/ncsgzdt/200704/t20070416_53742.htm.

71. Interview with an agricultural economist, Beijing, November 3, 2014; Cong Yuhua, "RMB20 Billion to Break the GM Crop Deadlock" (in Chinese), *China Youth Daily*, September 24, 2008.

72. Carl E. Pray et al., "The Impact of Bt Cotton and the Potential Impact of Biotechnology on Other Crops in China and India," *Frontiers of Economics and Globalization* 10 (2011): 85.

73. Annual accounts of China's Ministry of Agriculture (Beijing: Ministry of Agriculture, various years). I would like to thank Yutao Sun for locating this information.

74. Huang et al., "Plant Biotechnology in China"; Wang Dayuan, "China Is the First Country in the World to Commercialize GM Crops" (in Chinese), blog post, May 14, 2014, accessed June 22, 2014, http://blog.sciencenet.cn/blog-642008-794435.html.

75. Bureau of Rural and Social Development and China National Center for Biotechnology Development, under the Ministry of Science and Technology, the People's Republic of China, *China's Biotechnology Development Report* 2002 (in Chinese) (Beijing: China Agricultural Press, 2003), 217.

76. Li et al., "Biosafety Management and Commercial Use of Genetically Modified Crops in China," 569.

77. Interview with a biotechnology policy researcher, Beijing, November 23, 2002.

78. An "event" is defined as a specific set of genes that have been placed in a specific plant background material. Thus, many Bt-cotton events consist of different types of Bts in one background or the same Bts in different backgrounds. Pray et al., "The Impact of Bt Cotton and the Potential Impact of Biotechnology on Other Crops in China and India," 86.

79. Li et al., "Biosafety Management and Commercial Use of Genetically Modified Crops in China," 570.

80. Huang et al., *Economic Impacts of Transgenic Biotechnology*.

81. Office of Agricultural Genetic Engineering Biosafety Administration under the Ministry of Agriculture, Institute of Biotechnology under the Chinese Academy of Agricultural Sciences, and China Society for Agricultural Biotechnology, comps., *Thirty Years of Transgenic Technology* (in Chinese) (Beijing: China Agricultural Science and Technology Press, 2012).

82. This information was provided by Lijun Liu, which is much appreciated.

83. Interview with an agricultural biotechnologist, Beijing, November 10, 2010.

84. Tao Zhang and Shudong Zhou, "The Economic and Social Impact of GMOs in China," *China Perspectives* no. 47 (June 2003): 50–57; Yanqing Wang and Sam Johnston, "The Status of GM Rice R&D in China," *Nature Biotechnology* 25, no. 7 (2007): 717–718.

85. Huang et al., "Plant Biotechnology in China," 675.

86. Huang et al., *Economic Impacts of Transgenic Biotechnology*, 38.

87. Interview with an agricultural biotechnologist, Beijing, November 10, 2010, and an agricultural policy analyst, Beijing, November 17, 2010.

88. Huang et al., *Economic Impacts of Transgenic Biotechnology*, 39.

89. Interview with an agricultural biotechnologist, Beijing, August 30, 2011.

90. Jerry McBeath and Jenifer Huang McBeath, "The Socio-Political Environment of Biosafety Regulation in China: The Case of GMOs," in *Risk Assessment and Regulation of Genetically Modified Organisms: Proceeding of the Fourth International Biosafety Workshop*, ed. Dayuan Xue (Beijing: China Environmental Science Press, 2012), 104–134.

91. Chen, "Feeding the Nation," 85.

92. Peter Ho, Jennifer H. Zhao, and Dayuan Xue, "Access and Control of Agro-biotechnology: Bt Cotton, Ecological Change and Risk in China" (in Chinese), in *Risk Assessment and Safety Regulation of Genetically Modified Organisms: Proceedings of the International Biosafety Forum—Workshop 3*, ed. Dayuan Xue, Yoke Ling Chee, and Kun Xue (Beijing: China Environmental Science Press, 2009), 180.

93. Huang et al., "Plant Biotechnology in China," 675.

94. Nao Nakanishi, "China Seen a Crouching Dragon in Biotechnology," *Reuters*, December 20, 2002.

95. Clive James, *ISAAA Brief 51: 20th Anniversary of the Global Commercialization of Biotech Crops (1996 to 2015) and Biotech Crop Highlights in 2015*, Executive Summary, accessed June 1, 2016, http://www.isaaa.org/resources/publications/briefs/51/executivesummary/default.asp; ISAAA, *Brief 52: Global Status of Commercialized Biotech/GM Crops: 2016*, Executive Summary, accessed June 10, 2017, http://www.isaaa.org/resources/publications/briefs/52/executivesummary/default.asp.

96. Robert Tripp, "Transgenic Cotton and Institutional Performance," in *Biotechnology and Agricultural Development: Transgenic Cotton, Rural Institutions and Resource-Poor Farmers*, ed. Robert Tripp (London: Routledge, 2009), 89; Huang et al., "Farmers' Seed and Pest Control Management for Bt Cotton in China," *Biotechnology and Agricultural Development: Transgenic Cotton, Rural Institutions and Resource-Poor Farmers*, ed. Robert Tripp (London: Routledge, 2009), 110.

97. Tripp, "Transgenic Cotton and Institutional Performance," 110.

98. Huang et al., *Economic Impacts of Transgenic Biotechnology*, 133–134.

99. Interview with a staff member at Monsanto China, Beijing, November 26, 2010.

100. Carl E. Pray et al., "Impact of Bt Cotton in China," *World Development* 29 (2001): 813–825.

101. Jikun Huang et al., "Biotechnology Boosts to Crop Productivity in China: Trade and Welfare Implications," *Journal of Development Economics* 75, no. 1 (2004): 27–54.

102. In India, activists feared that the introduction of Bt cotton would impoverish farmers who would be forced into buying expensive seeds from multinational companies and go further into debt. They have argued that the introduction of Bt cotton will therefore result in the economic colonization of India by outsiders. For a discussion on the policy difference between China and India regarding Bt cotton, see Bhagirath Choudhary, "Development of Transgenic Bt Cotton Technology in India and China: A Policy Perspective," *Science and Public Policy* 28, no. 3 (2001): 219–229.

103. Pray et al., "The Impact of Bt Cotton and the Potential Impact of Biotechnology on Other Crops in China and India," 92–97.

104. Pray et al., "The Impact of Bt Cotton and the Potential Impact of Biotechnology on Other Crops in China and India," 92–97.

105. Jikun Huang, et al., "Biotechnology as an Alternative to Chemical Pesticides: A Case Study of Bt Cotton in China," *Agricultural Economics* 29, no. 1 (2003): 55–67.

106. Kong-Ming Wu et al., "Suppression of Cotton Bollworm in Multiple Crops in China in Areas with Bt Toxin–Containing Cotton," *Science* 321 (September 19, 2008): 1676–1678.

107. Pray et al., "The Impact of Bt Cotton and the Potential Impact of Biotechnology on Other Crops in China and India," 92–97.

108. Kym Anderson and Shunli Yao, "China, GMOs and World Trade in Agricultural and Textile Products," *Pacific Economic Review* 8, no. 2 (2003): 167–168.

109. Cong, "RMB20 Billion to Break the GM Crop Deadlock"; Huang et al., "Plant Biotechnology in China"; Huang et al., *Economic Impacts of Transgenic Biotechnology*.

    According to Huang Dafang at the CAAS Institute of Biotechnology, Bt cotton creates an economic benefit of RMB8 billion annually, which has cumulated to more than RMB40 billion over the years. See "GM Rice Have Waited Five Years for Their Biosafety Certificates" (in Chinese), *People's Daily* Online live interview transcript, December 25, 2009, accessed January 11, 2010, http://live.people.com.cn/note.php?id=662091225085841_ctdzb_008.

110. Denis Fred Simon and Cong Cao, *China's Emerging Technological Edge: Assessing the Role of High-End Talent* (Cambridge: Cambridge University Press, 2009), 212–253.

4. Science, Biosafety, and Regulations

1. Lester Ross and Walter Hutchens, "Genetic Modification in Agriculture: The Impact of China's Regulations on Foreign Trade and Investment," *China Law & Practice* 15, no. 9 (2001): 30–37; Mary A. Marchang, Cheng Fang, and Bao-hui Song, "Issues on Adoption, Import Regulations, and Policies for Biotech Commodities in China with a Focus on Soybeans," *AgBioForum* 5, no. 4 (2002): 167–174.

2. Nancy N. Chen, "Feeding the Nation: Chinese Biotechnology and Genetically Modified Food," in *Asian Biotech: Ethics and Communities of Fate*, ed. Aihwa Ong and Nancy N. Chen (Durham, NC: Duke University Press, 2010), 82.

3. Carl E. Pray et al., "Costs and Enforcement of Biosafety Regulations in India and China," *Journal of Technology and Globalisation* 2, no. 1/2 (2004): 144.

4. Pray et al., "Costs and Enforcement of Biosafety Regulations in India and China."

5. Jikun Huang et al., *Economic Impacts of Transgenic Biotechnology: Ten Years of Bt-Cotton in China* (in Chinese) (Beijing: Science Press, 2010), 27–30.

6. Zhang Qifa et al., "Suggestions on the Development Strategy of Research and Commercialization of GM Crops in China" (in Chinese), *Bulletin of the Chinese Academy of Sciences* 19, no. 5 (2004): 330–332.

7. Robert L. Paarlberg, *The Politics of Precaution: Genetically Modified Crops in Developing Countries* (Baltimore: Johns Hopkins University Press, 2001), 129.

8. Interview with a Greenpeace China campaigner, Beijing, October 29, 2010.

9. Interview with a researcher on biotechnology policy, Beijing, September 6, 2002.

10. Yanqing Wang and Sam Johnston, "The Status of GM Rice R&D in China," *Nature Biotechnology* 25, no. 7 (2007): 717–718.

11. The labeling requirement was initially to protect Chinese soybean farmers, who were happy with the labeling of foods produced with GM soybeans. Interview with a bioethics expert, Beijing, December 25, 2002.

12. Yu Wenxuan and Chen Yi, "China's Agro-GMO Biosafety Legislation and Its Improvement," in *Risk Assessment and Regulation of Genetically Modified Organisms: Proceedings of the Fourth Workshop of the International Biosafety Forum*, ed. Xue Dayuan (Beijing: China Environmental Science Press, 2012), 177–188.

13. Pray et al., "Costs and Enforcement of Biosafety Regulations in India and China," 145.

14. MOA also issued the "Biosafety Assessment of Plant GM Organisms and a National Standard for Monitoring the Environmental Safety of GM Plants and Relevant Products" in 2007 to impose biosafety regulations on all varieties generated from transgenic technology and on the sale of GM seeds.

15. Yunhe Li et al., "Biosafety Management and Commercial Use of Genetically Modified Crops in China," *Plant Cell Report* 33 (2014): 567; Jikun Huang and Qingfang Wang, "Biotechnology Policy and Regulation in China," IDS Working Paper 195 (Brighton, UK: University of Sussex, Institute of Development Studies, 2003).

16. Peter Oosterveer, *Global Governance of Food Production and Consumption: Issues and Challenges* (Cheltenham, UK: Edward Elgar, 2007), 137.

17. The following several paragraphs draw extensively from Jerry McBeath and Jenifer Huang McBeath, "The Socio-Political Environment of Biosafety Regulation in China: The Case of GMOs," in *Risk Assessment and Regulation of Genetically Modified Organisms: Proceedings of the Fourth Workshop of the International Biosafety Forum*, ed. Xue Dayuan (Beijing: China Environmental Science Press, 2012), 104–134.

18. Peter Newell, "Domesticating Global Policy on GMOs: Comparing India and China," IDS Working Paper 206 (Brighton, UK: University of Sussex, Institute for Development Studies, 2003).

19. Interview with a staff member of Monsanto China, Beijing, November 26, 2010.

20. For a discussion of the relationship between trade and GM foods, see Alasdair R. Young, "Political Transfer and 'Trading Up'? Transatlantic Trade in Genetically Modified Food and U.S. Politics," *World Politics* 55, no. 4 (2003): 457–484.

21. Interview with a bioethics expert, Beijing, December 25, 2002.
22. Zhu Rongji, "Develop Ourselves by Fully Utilizing the Opportunity of WTO Accession" (in Chinese), February 24, 2002, in *Zhu Rongji on the Record*, comp. Editorial Board (Beijing: People's Press, 2011), 4: 320–322.
23. Tao Zhang and Shudong Zhou, "The Economic and Social Impact of GMOs in China," *China Perspectives*, no. 47 (June 2003): 50–57.
    Similarly, in addition to religious reasons, Thailand's rice exports are very important to the country, and it cannot afford to have any country ban Thai rice out of concern that the rice has been genetically modified.
24. Ron Herring, "China, Rice, and GMOs: Navigating the Global Rift on Genetic Engineering," *Asia-Pacific Journal* 7, no. 3 (January 2009), accessed January 16, 2009, http://japanfocus.org/-Ron-Herring/3012.
25. Office of Agricultural Genetic Engineering Biosafety Administration Under the Ministry of Agriculture, Institute of Biotechnology Under the Chinese Academy of Agricultural Sciences, and China Society for Agricultural Biotechnology, comps., *Thirty Years of Transgenic Technology* (in Chinese) (Beijing: China Agricultural Science and Technology Press, 2012), 23.
26. Li et al., "Biosafety Management and Commercial Use of Genetically Modified Crops in China"; Fu Zhongwen, "The Progress on Biosafety Regulation of Agri-GMOs in China" (in Chinese), in *Risk Assessment and Management of Genetically Modified Organisms: Proceedings of the Fifth Workshop of the International Biosafety Forum*, ed. Xue Dayuan (Beijing: China Environmental Science Press, 2012), 39–45.
27. Li et al., "Biosafety Management and Commercial Use of Genetically Modified Crops in China."
28. However, SEPA "has little input into the design of biosafety regulations, or decision-making processes in relation to particular GM product applications," and "SEPA will never say they are against GM, the aim is to put a safety straightjacket on biotech." See James Keeley, "Regulating Biotechnology in China: The Politics of Biosafety," IDS Working Paper 208 (Brighton, UK: University of Sussex, Institute of Development Studies, 2003), 8, 9.
29. Interview with a Greenpeace China campaigner, Beijing, October 29, 2010.
30. Interview with a bioethics expert, Beijing, December 25, 2002.
31. Interview with a biodiversity expert, Beijing, November 8, 2010.
32. Pray et al., "Costs and Enforcement of Biosafety Regulations in India and China"; Li et al., "Biosafety Management and Commercial Use of Genetically Modified Crops in China."
33. Several sources, including one from MOA, indicate that the first NBCAGMO had fifty-eight members, but only a list of fifty-seven members could be located. See, for example, "MOA: The National GMO Biosafety Committee Established" (in Chinese), accessed June 15, 2014, http://www.china.com.cn/chinese/PI-c/170495.htm. The following analysis is based on the number as fifty-seven.

34. Interview with an agricultural economist, Beijing, November 3, 2010.

35. An Baijie, "Court Rejects Lawsuit Over GM Rice: Lawyer," *Global Times*, March 26, 2010.

36. Zhang Wei, "'Covert' GMOs" (in Chinese), *China Youth Daily*, December 23, 2009; Keeley, "Regulating Biotechnology in China," 6–7.

37. Liu Jianqiang, "GM Rice Surprisingly Found on Markets" (in Chinese), *Southern Weekend*, April 14, 2005; interview with an agricultural biotechnologist, Beijing, November 10, 2010.

38. Cui Yongyuan, "Objection to the Composition of the Fifth NBCAGMO" (in Chinese), Weibo post, June 16, 2016, accessed December 27, 2016, http://weibo.com/ttarticle/p/show?id=2309403987131688203084.

39. Sheila Jasanoff, *The Fifth Branch: Science Advisers as Policymakers* (Cambridge, MA: Harvard University Press, 1998), 167.

40. I am indebted to Lijun Liu for suggesting the way to analyze NBCAGMO members according to their disciplinaries as well as their locations along the research and development value chain.

41. Interview, Beijing, November 8, 2010.

42. "GM Rice Have Waited Five Years for Their Biosafety Certificates" (in Chinese), *People's Daily* Online live interview transcript, December 25, 2009, accessed January 11, 2010, http://live.people.com.cn/note.php?id=662091225085841_ctdzb_008.

43. Huang et al., *Economic Impacts of Transgenic Biotechnology*, 29.

44. He Jun, "China Made GM Rice Quietly Granted Biosafety Certificates" (in Chinese), *Daily Economic News*, December 3, 2012.

45. Clive James, "China Approves Biotech Rice and Maize in Landmark Decision," *Crop Biotech Update*, December 4, 2009, accessed January 10, 2013, http://www.isaaa.org/kc/cropbiotechupdate/article/default.asp?ID=5112; see also Chen Weixiao, "China Made 'Landmark' GM Food Crop Approval," *SciDev.Net*, December 17, 2009, accessed January 10, 2013, http://www.scidev.net/global/biotechnology/news/china-makes-landmark-gm-food-crop-approval.html.

46. Jia Hepeng, "GM Rice May Soon Be Commercialized," *China Daily*, January 27, 2005.

47. Liu Jianqiang, "GM Rice for the 1.3 Billion Chinese: A Game of Safety and Interests" (in Chinese), *Southern Weekend*, December 9, 2004.

48. Cong Yuhua, "RMB20 Billion to Break the GM Crop Deadlock" (in Chinese), *China Youth Daily*, September 24, 2008.

49. Wang and Johnston, "The Status of GM Rice R&D in China."

50. Mao Chen, Anthony Shelton, and Gong-yin Ye, "Insect-Resistant Genetically Modified Rice in China: From Research to Commercialization," *Annual Review of Entomology* 56 (2011): 88.

51. Wang and Johnston, "The Status of GM Rice R&D in China."
52. Tu is now a professor at Zhejiang University. Interview, Hangzhou, July 18, 2013.
53. Interview with a plant geneticist, Wuhan, August 29, 2012.
54. Liu Zhiwei, and Fan Jingqun, "Safety of GM Rice Verified from Different Aspects" (in Chinese), *Science and Technology Daily*, January 7, 2010.
55. Deng Li, "China Cautious Towards GM Rice Commercialization" (in Chinese), *Twenty-First Century Business Herald*, July 10, 2008.
56. Chen Hongxia and Zhang Xu, "Biosafety Certificates Close to Renewal While Controversy Around Commercialization of Food Staple Continues" (in Chinese), *Twenty-First Century Business Herald*, August 19, 2014.
57. Zhang Wei, "'Covert' GMOs" (in Chinese), *China Youth Daily*, December 23, 2009.
58. An, "Court Rejects Lawsuit Over GM Rice."
59. Zhang, "'Covert' GMOs"; Meng Dengke, "GM Rice Allowed to Plant: Bashful Policy Released Quietly" (in Chinese), *Southern Weekend*, December 10, 2009.
60. Tian Jiaojin, "A Hundred or So Scholars Petition and CPPCC Members Propose: Commercialization of GM Rice Under Scrutiny" (in Chinese), *Youth Times*, March 11, 2010, A3.
61. "GM Rice Have Waited Five Years for Their Biosafety Certificates."
62. Xu Haigen et al., "The Construction of Biosafety Clearing House of China and Its Proposal," in *Environmental Impacts and Safety Regulation of Genetically Modified Organisms: Proceedings of the International Biosafety Forum*, ed. Xue Dayuan (Beijing: China Environmental Science Press, 2006), 96–97.
63. See, for example, Elizabeth Economy, *The River Runs Black: The Environmental Challenge to China's Future*, 2nd ed. (Ithaca, NY: Cornell University Press, 2010), for a discussion on environmental governance.
64. Huang et al., *Economic Impacts of Transgenic Biotechnology*, 43.
65. Liu, "GM Rice Surprisingly Found on Markets"; Jane Qiu, "Is China Ready for GM Rice?," *Nature* 455 (October 16, 2008): 851.
66. McBeath and McBeath, "The Socio-Political Environment of Biosafety Regulation in China"; Jikun Huang et al., "Farmers' Seed and Pest Control Management for Bt Cotton in China," *Biotechnology and Agricultural Development: Transgenic Cotton, Rural Institutions and Resource-Poor Farmers*, ed. Robert Tripp (London: Routledge, 2009), 114–115.
67. Liu, "GM Rice Surprisingly Found on Markets."
68. Craig Simons, "Of Rice and Men," *Newsweek*, December 20, 2004, accessed June 15, 2014, http://www.newsweek.com/rice-and-men-123141.
69. Li Yanjie, "Greenpeace: Thirty Percent of Rice Samples in Wuhan Contains GM Ingredients" (in Chinese), *China Business Journal*, May 17, 2014.

70. CCTV, *News Probe: Tracing GM Rice*, July 26, 2014, accessed July 20, 2017, http://zhongyang13.com/cctv13/xinwendiaocha/20140727/10297.html.

71. Chen, "Feeding the Nation," 84.

72. Ron J. Herring, "Stealth Seeds: Bioproperty, Biosafety, Biopolitics?," *Journal of Development Studies* 43, no. 1 (2007): 130–157.

73. "EU Tightens Control of Chinese Rice Over GM Fears," *EU Business*, November 15, 2011, accessed June 7, 2014, http://www.eubusiness.com/news -eu/china-biotech-food.diq.

74. Interview with a staff member at Monsanto China, Beijing, November 26, 2010.

75. Interview with a biodiversity expert, Beijing, December 27, 2002.

76. "EU Tightens Control of Chinese Rice Over GM Fears."

77. Xueman Wang, "Challenges and Dilemmas in Developing China's National Biosafety Framework," *Journal of World Trade* 38, no. 5 (2004): 899–913.

78. Keeley, "Regulating Biotechnology in China," 5; Huang and Wang, "Biotechnology Policy and Regulation in China."

79. Wei Guihong, "Research on Legislation Issue and Advice for GMOs-Safety in China" (in Chinese), in *Risk Assessment and Regulation of Genetically Modified Organism: Proceedings of the Fourth Workshop of the International Biosafety*, ed. Xue Dayuan (Beijing: China Environmental Science Press, 2012), 190, 192.

80. At one time, the biosafety law was expected to come out in October 2003. Keeley, "Regulating Biotechnology in China," 8–9.

81. Interview with a Greenpeace China campaigner, Beijing, October 29, 2010; interview with an expert of biodiversity, Beijing, November 8, 2010.

82. Interview with a science journalist, Beijing, November 8, 2010.

83. Harvey Brooks, "The Scientific Adviser," in *Scientists and National Policy-Making*, ed. Robert Gilpin and Christopher Wright (New York: Columbia University Press, 1964), 76–77.

5. Polarization and Politicization of Transgenic Technology

1. Ray Offenheiser and Kimberly Pfeifer, "Forward," in *Biotechnology and Agricultural Development: Transgenic Cotton, Rural Institutions and Resource-Poor Farmers*, ed. Robert Tripp (New York: Routledge, 2009), xxi.

2. Greenpeace, "Chinese Consumer Loses GE Food Labelling Case Against Nestlé but Vows to Continue," April 20, 2004, accessed October 10, 2014, http://www.greenpeace.org/eastasia/news/stories/food-agriculture/2004 /chinese-consumer-loses-ge-food/.

3. Greenpeace, "Chinese Consumer Loses GE Food Labelling Case."

4. Yu Jingyin, "It's Safe to Eat Imported GM Soybean Oils" (in Chinese), *People's Daily* (Overseas Edition), June 21, 2013.

5. Yu Dawei, "GM Rice: The Eve of Its Commercialization" (in Chinese), *New Century Weekly*, no. 8 (February 22, 2010), accessed February 17, 2010, http://magazine.caixin.com/2010-02-17/100117846.html.

6. Yanqing Wang and Sam Johnston, "The Status of GM Rice R&D in China," *Nature Biotechnology* 25, no. 7 (2007): 717–718.

7. Huangguang Qiu et al., "Consumers' Trust in Government and Their Attitudes Towards Genetically Modified Food: Empirical Evidence from China," *Journal of Chinese Economic and Business Studies* 10, no. 1 (2012): 67–87; interview with a science journalist, Beijing, November 8, 2010.

8. "Ideological Debate in China: The Little Red Bookstore," *Economist*, February 5, 2009. The website was shut down in 2012, amid the crackdown on Bo Xilai, Chongqing's former party secretary, who was expelled from the party, found guilty of corruption, and sentenced to life imprisonment. But it revived later under a different address. Tania Branigan, "China Shut Down Maoist Website Utopia," *Guardian*, April 6, 2012; interview with a science journalist, Beijing, October 29, 2010.

9. Tian Jiaojin, "A Hundred or So Scholars Petition and CPPCC Members Propose: Commercialization of GM Rice Under Scrutiny" (in Chinese), *Youth Times*, March 11, 2010, A3.

10. Tian, "A Hundred or So Scholars Petition and CPPCC Members Propose."

11. Jiang Jingsong et al., "An Appeal to Postpone the Commercialization of GM Staple Foods" (in Chinese), blog post, March 11, 2011, accessed November 10, 2014, http://blog.sciencenet.cn/home.php?mod=space&uid=224810&do=blog&id=301777.

12. Richard Stone, "Activists Go on Warpath Against Transgenic Crops—and Scientists," *Science* 331 (February 25, 2011): 1000–1001.

    However, according to an account by a GMO activist, Zhang Qifa was unable to answer three questions raised by a young person and fled in panic. Zhang also had to cancel another lecture in December 2010. See Yin Shuai-Jun, "Social Risk Management and Control for GMOs" (in Chinese), in *Risk Assessment and Risk Management of the Genetically Modified Organisms: Proceedings of the Fifth Workshop of the International Biosafety*, ed. Xue Dayuan (Beijing: China Environmental Press, 2014), 109.

13. Chen Ming et al., "Fang's Method: Fang Zhouzi and the Debate Rules Under His Influence" (in Chinese), *Southern Weekend*, June 22, 2012. "Method" (*fangfa*) is homophonic to "Fang's method" (*fang fa*) in Chinese.

14. Jie Chen, *Transnational Civil Society in China: Intrusion and Impact* (Cheltenham, UK: Edward Elgar, 2012), 71–72; Lin Gu, "Greenpeace Landed in China Peacefully" (in Chinese), *Oriental Outlook Weekly* no. 4 (December 11, 2003): 26–27.

15. Interview with a Greenpeace China campaigner, Beijing, October 29, 2010; interview with an agricultural biotechnologist, Beijing, November 10, 2010.

16. Dayuan Xue, *A Summary of Research on the Environmental Impacts of Bt Cotton in China* (Hong Kong: Greenpeace, 2002).

17. James Keeley, "Regulating Biotechnology in China: The Politics of Biosafety," IDS Working Paper 208 (Brighton, UK: University of Sussex, Institute of Development Studies, 2003), 22–27; interview with an agricultural economist, Beijing, November 3, 2010.

18. Carl E. Pray et al., "The Impact of Bt Cotton and the Potential Impact of Biotechnology on Other Crops in China and India," *Frontiers of Economics and Globalization* 10 (2011): 107.

19. Xia Shang, "Science Communication Behaviors of Greenpeace China on Genetically Modified Organism Issues" (in Chinese) (master's thesis, Peking University, Beijing, 2007).

20. Liu Jianqiang, "GM Rice for the 1.3 Billion Chinese: A Game of Safety and Interests" (in Chinese), *Southern Weekend*, December 9, 2004.

21. Zhao Zuo, "Monsanto Phantom: Crisis on Chinese's Dining Table" (in Chinese), *Time Weekly*, July 8, 2010.

22. This was confirmed in an interview with an agricultural biotechnologist, Beijing, November 10, 2010.

23. Interview with a Greenpeace China campaigner, Beijing, October 29, 2010; interview with an agricultural biotechnologist, Beijing, November 10, 2010.

24. Peng Kefeng and Wang Shan, "Greenpeace's Rice Theft at Night" (in Chinese), *China Science Daily*, May 9, 2014.

25. Monica Tan, "24 Children Used as Guinea Pigs in Genetically Engineered 'Golden Rice' Trial," blog post, August 31, 2012, accessed May 31, 2014, http:// www.greenpeace.org/eastasia/news/blog/24-children-used-as-guinea-pigs -in-geneticall/blog/41956/.
    Much of the information in this section is drawn from Sun Tao, Cai Tingyi, and Xu Jin, "The Abnormal Golden Rice Experiment in Hengyang" (in Chinese), *Caijing Magazine*, no. 23 (September 10, 2012); Liang Wei, "The Hengyang Golden Rice Story" (in Chinese), *Time Weekly*, December 13, 2012.

26. Guangwen Tang et al., "β-Carotene in Golden Rice Is as Good as β-Carotene in Oil at Providing Vitamin A to Children," *American Journal of Clinical Nutrition* 96, no. 3 (2012): 658–664.

27. "Email of Zou Ping of MOA's Office of Agricultural Genetic Engineering Biosafety Administration to Lorena Luo of Greenpeace, July 24, 2008," in Wen Jiayun (the Third World Network), "MOA Called off Golden Rice Experiment in China in 2008" (in Chinese), blog post, August 31, 2012, accessed May 9, 2014, http://blog.sina.com.cn/s/blog_81b7cbe801014kt5.

html; "Project NCT00680212: Vitamin A Equivalence of Plant Carotenoids in Children," accessed May 9, 2014, http://clinicaltrials.gov/ct2/show/record /NCT00680212.

According to the website GM-Free Cymru, at one time information was available on the U.S. government clinical trial website indicating that the trial was approved by an ethics review board in Zhejiang on May 19, 2008; however, this approval was removed on July 16, 2008. This information was neither revealed in Tang's paper nor available when I last checked the clinical trial website on August 14, 2014.

Also according to the same website, the Tufts team carried out another trial with the same design and controls in 2004 and 2005 in Hangzhou, Zhejiang, in which twenty-four children (between seven and ten years old) were fed two meals of Golden Rice per day over a period of seven days (Project NCT 00082420: Retinol Equivalence of Plant Carotenoids in Children). Such information was also unavailable when I last checked the website on August 14, 2014. See Brian John, "Golden Rice Feeding Trials Breach Medical Ethics Code," blog post, no date, accessed August 14, 2014, http://www.gmfreecymru.org/pivotal_papers /feedingrice.html.

28. Martin Enserink, "Golden Rice Not So Golden for Tufts," *Science*, September 18, 2013, accessed May 10, 2014, http://news.sciencemag.org/asiapacific/2013/09 /golden-rice-not-so-golden-tufts.

29. Enserink, "Golden Rice Not So Golden for Tufts."

30. Cat Ferguson, "Rice Researcher in Ethics Scrape Threatens Journal with Lawsuit Over Coming Retraction," *Retraction Watch*, July 17, 2014, accessed August 10, 2014, http://retractionwatch.com/2014/07/17/rice-researcher-in -ethics-scrape-threatens-journal-with-lawsuit-over-coming-retraction/; Alison McCook "Golden Rice Paper Pulled After Judge Rules for Journal," *Retraction Watch*, July 30, 2015, accessed August 30, 2015, http://retractionwatch .com/2015/07/30/golden-rice-paper-pulled-after-judge-rules-for-journal/.

31. Enserink, "Golden Rice Not So Golden for Tufts."

32. Dan Charles, "Golden Rice Study Violated Ethical Rules, Tufts Says," *National Public Radio*, September 17, 2013, accessed October 30, 2014, http:// www.npr.org/blogs/thesalt/2013/09/17/223382375/golden-rice-study -violated-ethical-rules-tufts-says.

33. Jane Qiu, "China Sacks Officials Over Golden Rice Controversy," *Nature*, December 10, 2012, accessed January 10, 2013, http://www.nature.com/news /china-sacks-officials-over-golden-rice-controversy-1.11998.

34. Jeffrey M. Smith, *Seeds of Deception: Exposing Industry and Government Lies About the Safety of the Genetically Engineered Foods You're Eating*, 3rd ed. (Fairfield, IA: Yes! Books, 2003); *Genetic Roulette: The Documented Health Risks of Genetically Engineered Foods*, 4th ed. (Fairfield, IA: Yes! Books, 2007).

35. F. William Engdahl, *Seeds of Destruction: The Hidden Agenda of Genetic Manipulation* (Montreal: Global Research, 2007).

36. Marie-Monique Robin, *The World According to Monsanto: Pollution, Corruption, and the Control of Our Food Supply*, trans. George Holoch (New York: New Press, 2010).

37. F. William Engdahl, "About F. William Engdahl—Biography," accessed August 16, 2016, http://www.williamengdahl.com/about.php.

38. In fact, Engdahl's other books—*Full Spectrum Dominance: Totalitarian Democracy in the New World Order* (2009), *A Century of War: Anglo-American Oil Politics and the New World Order* (2011), *Gods of Money: Wall Street and the Death of the American Century* (2011), *Myths, Lies and Oil Wars* (2012), and *Target: China—How Washington and Wall Street Plan to Cage the Asian Dragon* (2014)—all have Chinese translations. China's Intellectual Property Publishing House has included translations of four books—*Food Crisis, Financial Tsunami, American Hegemony*, and *Petroleum War*—in a geopolitics series. The Democracy and Rule of Law Press, another publisher that used to be affiliated with the General Office of the National People's Congress, has published translations of his two other books—*Targeting China: Washington's Dragon-Killing Strategy* and *The New Energy Wars: The Next Target China*. Some of the books may have been published with different titles. Books by Jeffrey M. Smith and Marie-Monique Robin also have been translated into Chinese.

39. India's GMO activist Vandana Shiva has also found her audience in China, where her books have been translated into Chinese and she has been welcomed and provided a platform to challenge Monsanto and other agricultural biotechnology MNCs.

40. F. William Engdahl, "Engdahl China Speaking Tour (6)" (in Chinese), blog post, November 4, 2008, accessed June 22, 2014, http://blog.sina.com.cn/s/blog_5d21454b0100b8ka.html.

41. Engdahl, *Seeds of Destruction*, 127.

42. Gu Xiulin, *GMO War: Defending China's Food Security in the Twenty-First Century* (in Chinese) (Beijing: Intellectual Property Publishing House, 2011).

43. Gu Xiulin, Weibo post, accessed August 16, 2016, http://blog.sina.com.cn/gxiulin.

44. Zhang Weibin, "Who Defeat Scientists on GMOs?" (in Chinese), *Hubei Daily*, December 2, 2013.

45. Cui Zheng et al., "Monsanto's Cautious China Strategy," *Caixin Online*, December 24, 2013, accessed September 25, 2014, http://english.caixin.com/2013-12-24/100621301.html.

46. "Key Figures in the Development of GM Staple Foods Are Implicitly Connected to the Rockefeller Foundation" (in Chinese), *Business Week*, March 15, 2010, 15.

47. Lü Yongyan, "Netizens Reveal That Wang Dafang Is a Spy of the GMO Group" (in Chinese), blog post, March 12, 2012, accessed June 22, 2014, http://blog.sina.com.cn/s/blog_51508bd70102duyp.html.

48. Liu Yi, "The Vice Minister Supporting GMOs Accused to Be Employed by an American Company" (in Chinese), *Beijing Youth Daily*, October 31, 2013.

49. The story came up in an interview with a plant ecologist, Beijing, November 15, 2010. It resurfaced in mid-2017 when Sun was removed for corruption from his position as Chongqing's party secretary.

50. Song Yangbiao, "The Debates on GM Rice as the Staple Food Up Again" (in Chinese), *Time Weekly*, March 11, 2010.

51. Ji Tianqin et al., "Scholar Wang Lijun" (in Chinese), *Southern Cosmopolitan Weekly* no. 48 (December 28, 2012).

52. Peng Guangqian, "Expert's Eight Questions on the Transgenization of Staple Foods: Why Should China Introduce the Technology Blindly" (in Chinese), *Global Times*, August 21, 2013. There is only a one-character difference between "pie" (*xianbing*) and "trap" (*xianjing*) in Chinese.

53. Robert K. Merton, "The Normative Structure of Science," in *The Sociology of Science: Theoretical and Empirical Investigations*, ed. Norman W. Store (Chicago: University of Chicago Press, 1973 [1942]), 267–278.

54. Wang Can, "Qin Zhongda, the 92-Year-Old Former Minister of Chemical Industry and a Former Member of the CCP's Central Committee, Questioned China's Biggest Overseas Acquisition" (in Chinese), *The Paper* (an online news site), April 12, 2016, accessed August 16, 2016, http://www.thepaper.cn/newsDetail_forward_1453446.

55. Liu Jianqiang, "GM Rice for the 1.3 Billion Chinese." What Lenin really said was "A dozen wise men can be more easily wiped out than a hundred fools."

56. Interview with an agricultural policy analyst, Beijing, April 6, 2014.

57. Yan Dingfei and Huang Boxin, "Thousand Persons Try GM Rice: Is It a Serious Activity of Science Popularization" (in Chinese), *Southern Weekend*, July 26, 2013.

Tasting unregulated food is not just a Chinese phenomenon. On October 20, 2016, Cellectis, one of the companies developing gene-edited crops, hosted a dinner at Benoit New York, an upscale French restaurant opened by Alain Ducasse, a renowned chef, serving dishes made from its gene-edited soybeans and potatoes to guests including professors, journalists, and celebrities like actor Neil Patrick Harris. Kenneth Chang, "These Foods Aren't Genetically Modified but They Are 'Edited,' " *New York Times*, January 10, 2017.

58. Fang Zhouzi, "This Shows the Power of Science, Remark in Beijing on the Occasion of the National GM Rice Tasting Session" (in Chinese), blog post, August 17, 2014, accessed August 18, 2014, http://fangzhouzi.baijia.baidu.com/article/26328.

59. Interview with an agricultural policy analyst, Beijing: April 6, 2014.

60. Tom Phillips, "China Passes Law Imposing Security Controls on Foreign NGOs," *Guardian*, April 28, 2016.

61. See, for example, Guobin Yang and Craig Calhoun, "Media, Civil Society, and the Rise of a Green Public Sphere in China," *China Information* 21, no. 2 (2007): 211–236; and Andrew C. Mertha, *China's Water Warriors: Citizen Action and Policy Change* (Ithaca, NY: Cornell University Press, 2008).

62. Colin Macilwain, "Against the Grain," *Nature* 422 (March 13, 2003): 111–112.

63. C. Wright Mills, *The Power Elite* (Oxford: Oxford University Press, 1956).

64. David Vogel, *The Politics of Precaution: Regulating Health, Safety, and Environmental Risks in Europe and the United States* (Princeton, NJ: Princeton University Press, 2012), 37.

6. The Chinese Media and Changing Policy

Yuxian Liu provided substantial assistance in the preparation of the chapter, which is greatly appreciated.

1. Michael Gurevitch and Mark R. Levy, eds., *Mass Communication Review Yearbook*, vol. 5 (Thousand Oaks, CA: Sage, 1986), 19, quoted in William L. Gamson and Andre Modigliani, "Media Discourse and Public Opinion on Nuclear Power: A Constructivist Approach," *American Journal of Sociology* 95, no. 1 (July 1989): 1–37.

2. Susanna Hornig, "Reading Risk: Public Response to Print Media Accounts of Technological Risk," *Public Understanding of Science* 2, no. 2 (April 1993): 95–109; Gaye Tuchman, "Telling Stories," *Journal of Communication* 26, no. 4 (December 1976): 93–97.

3. Paul D'Angelo, "News Framing as a Multiparadigmatic Research Program: A Response to Entman," *Journal of Communication* 52, no. 4 (December 2002), 870–871.

4. Dorothy Nelkin, "The Culture of Science Journalism," *Society* 24, no. 6 (September/October 1987): 17–25.

5. Martin W. Bauer, "Controversial Medical and Agri-Food Biotechnology: A Cultivation Analysis," *Public Understanding of Science* 22, no. 2 (2002): 93–111.

6. Leonie A. Marks et al., "Mass Media Framing of Biotechnology News," *Public Understanding of Science* 16, no. 2 (2007): 183–203.

7. Guo Yuhua, "Angel or Evil? Social and Cultural Perspective to GM Soybeans in China" (in Chinese), *Sociological Study*, no. 1 (2005): 84–112.

8. Li Du and Christen Rachul, "Chinese Newspaper Coverage of Genetically Modified Organisms," *BMC Public Health* 12 (2012): 326; Zhao Juan, "News Coverage of GM Crops in China: Problems and Solutions" (in Chinese), *Forum of Chinese Association of Science and Technology*, no. 2 (2012): 191–192.

9. Elisabeth Rosenthal, "Under Pressure, Chinese Newspaper Pulls Exposé on a Charity," *New York Times*, March 24, 2002.

10. Yuan Yue, *Analysis of News of Genetically Modified Crops and Foods in Four Chinese Newspapers* (in Chinese) (master's thesis, Graduate School of the Chinese Academy of Sciences, Beijing, 2008), 19–20.

11. Leonie A. Marks et al., "Media Coverage of Agrobiotechnology: Did the Butterfly Have an Effect?," *Journal of Agribusiness* 21, no. 1 (Spring 2003): 1–20.

12. Du and Rachul, "Chinese Newspaper Coverage of Genetically Modified Organisms."

13. Li Min, *A Case Study of the Coverage of GM Foods and Crops in People's Daily from the International Perspective* (master's thesis, Tsinghua University, Beijing, 2007), quoted in Qu Lulu, *Reporting Risks on GMOs: An Analysis of the 15-Year News of Genetic Risks in Science and Technology Daily* (in Chinese) (master's thesis, Graduate School of the Chinese Academy of Sciences, Beijing, 2009), 68–69.

14. Du and Rachul, "Chinese Newspaper Coverage of Genetically Modified Organisms."

15. Julie Leask, Claire Hooker, and Catherine King, "Media Coverage of Health Issues and How to Work More Effectively with Journalists: A Qualitative Study," *BMC Public Health* 10 (2010): 535.

16. Herbert J. Gans, *Deciding What's News: A Study of CBS Evening News, NBC Nightly News, Newsweek, and Time* (New York: Pantheon Book, 1979), 173–176.

17. Nelkin, "The Culture of Science Journalism"; Neal E. Miller, "The Scientist's Responsibility for Public Information: A Guide to Effective Communication with the Media," *Scientists and Journalists: Reporting Science as News*, ed. Sharon M. Freidman, Sharon Dunwoody, and Carol L. Rogers (New York: Free Press, 1986), 239–253.

18. Dominique Brossard and James Shanahan, "Do Citizens Want to Have Their Say? Media, Agricultural Biotechnology, and Authoritarian Views of Democratic Process in Science," *Mass Communication & Society* 6, no. 3 (2003): 306.

19. Matthew C. Nisbet and Bruce V. Lewenstein, "Biotechnology and the American Media: The Policy Process and the Elite Press, 1970 to 1999," *Science Communication* 23, no. 4 (June 2002): 359–391.

20. Sharon Dunwoody, "The Science Writing Inner Club: A Communication Link Between Science and the Lay Public," in *Scientists and Journalists: Reporting Science as News*, ed. Sharon M. Friedman, Sharon Dunwoody, and Carol L. Rogers (New York: Free Press, 1986), 155–169.

21. Dorothy Nelkin, "Selling Science," *Physics Today* 43, no. 11 (1990): 41–46.

22. Martin W. Bauer, "The Evolution of Public Understanding of Science—Discourse and Comparative Evidence," *Science, Technology & Society* 14, no. 2 (2009): 221–240.

23. Nelkin, "The Culture of Science Journalism."

24. Sharon M. Freidman, Sharon Dunwoody, and Carol L. Rogers, eds., *Scientists and Journalists: Reporting Science as News* (New York: Free Press, 1986).

25. Interview with a science journalist, Beijing, April 24, 2014.

26. Li Min, *A Case Study of the Coverage of GM Foods and Crops in People's Daily from the International Perspective*, quoted in Qu Lulu, *Reporting Risks on GMOs*, 68–69.

27. Nisbet and Lewenstein, "Biotechnology and the American Media," 385–386.

28. Du and Rachul, "Chinese Newspaper Coverage of Genetically Modified Organisms."

29. John R. Zaller, *The Nature and Origins of Mass Opinion* (Cambridge: Cambridge University Press, 1992).

30. Jan M. Gutteling et al., "Media Coverage 1973–1996: Trends and Dynamics," in *Biotechnology: The Making of a Global Controversy*, ed. Martin W. Bauer and George Gaskell (Cambridge: Cambridge University Press, 2002), 95–128.

31. Funing Zhong et al., "GM Food: A Nanjing Case Study of Chinese Consumers' Awareness and Potential Attitude," *AgBioForum* 5, no. 4 (2003): 136–144.

32. Xia Shang, *Science Communication Behaviors of Greenpeace China on Genetically Modified Organism Issues* (in Chinese) (master's thesis, Peking University, Beijing, 2007); interview with a science journalist, Beijing, November 4, 2010.

33. Roger E. Kaspersonn et al., "The Social Amplification of Risk: A Conceptual Framework," *Risk Analysis* 8, no. 2 (1988): 177–187.

34. Yuan, *Analysis of News of Genetically Modified Crops and Foods in Four Chinese Newspapers*, 44–45.

35. Surveys of Italian public opinion have noted, "Although lack of information on biotechnologies and a marked hostility against food biotechnologies are clear, the links between media exposure, levels of awareness, and attitudes toward biotechnologies are far from straightforward." See Massimiano Bucchi and Federico Neresini, "Why Are People Hostile to Biotechnologies?," *Science* 304 (June 18, 2004): 1749.

36. Zhang and Zhou, "The Economic and Social Impact of GMOs in China."

37. Zhong et al., "GM Food: A Nanjing Case Study of Chinese Consumers Awareness and Potential Attitude."

38. Brossard and Shanahan, "Do Citizens Want to Have Their Say?," 306.

39. Richard Stone, "Activists Go on Warpath Against Transgenic Crops—and Scientists," *Science* 331 (February 25, 2011): 1000–1001.

40. Mao Lei and Liao Wengen, "Wu Baoguo Chairs the NPC Standing Committee's Sixteenth Seminar" (in Chinese), *People's Daily*, June 26, 2010: 3.

41. Liu Jianqiang, "GM Rice for the 1.3 Billion Chinese: A Game of Safety and Interests" (in Chinese), *Southern Weekend*, December 9, 2004.

42. Meng Dengke, "GM Rice Allowed to Plant: Bashful Policy Released Quietly" (in Chinese), *Southern Weekend*, December 10, 2009.

43. Peng Liguo and Li Xudong, "Search for the Truth of Safety of Transgene in China: The Hidden Secrets" (in Chinese), *Southern Weekend*, May 13, 2011; Liu Jianqiang, "GM Rice Surprisingly Appears on the Market" (in Chinese), *Southern Weekend*, April 4, 2005.

44. Ke Bei, "The Ignorance and Prejudice toward GMOs" (in Chinese), *Southern Weekend*, July 22, 2011.

45. Li Tie, "Analysis of Chinese-Style Fallacy and Rumors about GMOs" (in Chinese), *Southern Weekend*, August 19, 2011.

46. "GM Rice Cultivation: Why Should China Aim for the First" (in Chinese), *Outlook Weekly*, December 7, 2009.

47. "Doubts About GM Rice" (in Chinese), *Outlook Weekly*, February 8, 2010.

48. Yuqiong Zhou and Patricia Moy, "Parsing Framing Processes: The Interplay Between Online Public Opinion and Media Coverage," *Journal of Communication* 57, no. 1 (2007): 79–98.

49. Zhang Zhongfa, "Commercialization and Safety of Transgenic Biotechnology" (in Chinese), in *China Development Studies 2002: Selected Research Reports of Development Research Center of the State Council*, ed. Ma Hong and Wang Mengkui (Beijing: China Development Press, 2002), 520.

    In the early twenty-first century, two Beijing newspapers that published GMO stories were criticized by CPPCC's Propaganda Department. Interview with a newspaper editor, Beijing, December 4, 2003.

50. Remarks by Xue Dayuan, a biodiversity expert, as quoted in Peter Ho and Eduard B. Vermeer, "Food Safety Concerns and Biotechnology: Consumers' Attitudes to Genetically Modified Products in Urban China," *AgBioForum* 7, no. 4 (2004): 169; interview with a bioethics expert, Beijing, December 25, 2002.

    Several science journalists interviewed in China indicated that except for *People's Daily* and Xinhua News Agency, the government is less likely to restrict reporting on GMOs; interviews, Beijing, April 22 and 24, 2014.

51. Interview with a science journalist, Beijing, April 24, 2014.

52. Interview with a science journalist, Beijing, April 21, 2014.

53. "Confronting the Bad News Bearers," *MIT Technology Review* 95, no. 7 (October 1992): 20–31.

54. Interview, Beijing, November 3, 2010.

55. Interview with a science journalist, Beijing, April 21, 2014; interview with an agricultural economist, Beijing, August 29, 2011; interview with an agricultural policy analyst, Beijing, August 30, 2011.

56. Brossard and Shanahan, "Do Citizens Want to Have Their Say?," 294; interviews with two science journalists, Beijing: October 29, 2010 and April 24, 2014 and a transgenic scientist, Beijing: November 8, 2010.

57. Jin Wei and Yu Shengnan, "Rats Disappeared: An Investigation Into Abnormal Animal Phenomenon in Shanxi and Jilin" (in Chinese), *International Herald Leader*, September 21, 2010; Jin Wei, "XY335: Rapid Expansion of an American Seed" (in Chinese), *International Herald Leader*, September 21, 2010; Zhang Luyu, "Editor's Note: Closing to the Truth Makes People So Worry" (in Chinese), *International Herald Leader*, September 21, 2010.

58. Jia Hepeng, Tan Yihong, and Fan Jingqun, *Handbook for Science Journalists Covering GMOs* (in Chinese) (Beijing: Center for Science Media, 2012).

59. George Gaskell and Martin W. Bauer, "Biotechnology in the Years of Controversy: A Social Scientific Perspective," in *Biotechnology 1996–1999: The Years of Controversy*, ed. George Gaskell and Martin W. Bauer (London: Science Museum Press, 2001), 3–14.

60. Julie L. Andsager, "How Interest Groups Attempt to Shape Public Opinion with Competing News Frames," *Journalism & Mass Communication Quarterly* 77, no. 3 (2000): 572–592; Lynn J. Frewer, Susan Miles, and Roy Marsh, "The Media and Genetically Modified Foods: Evidence in Support of Social Amplification of Risk," *Risk Analysis* 22, no. 4 (2002): 701–711.

61. Pippa Norris, *A Virtuous Circle: Political Communication in Postindustrial Societies* (Cambridge: Cambridge University Press, 2000); James Shanahan, "Television Viewing and Adolescent Authoritarianism," *Journal of Adolescence* 18, no. 3 (June 1995): 271–288; James Shanahan, "Television and Authoritarianism: Exploring the Concept of Mainstreaming," *Political Communication* 15, no. 4 (1998): 483–495.

62. John Durant, Martin W. Bauer, and George Gaskell, eds., *Biotechnology in the Public Sphere: A European Sourcebook* (London: Science Museum, 1998), quoted in Marks et al., "Media Coverage of Agrobiotechnology," 1.

63. Brossard and Shanahan, "Do Citizens Want to Have Their Say?," 306.

64. Paul Slovic, "Trust, Emotions, Sex, Politics, and Science: Surveying the Risk-Assessment Battlefield," in *The Perception of Risk*, ed. Paul Slovic (Sterling, VA: Earthscan, 1997), 394.

65. Matthew C. Nisbet, Dominique Brossard, and Adrianne Kroepsch, "Framing Science: The Stem Cell Controversy in an Age of Press/Politics," *Press/Politics* 8, no. 2 (2003): 36–70.

66. Interview with a Greenpeace China campaigner, Beijing, October 29, 2010; interview with an agricultural economist, Beijing, August 29, 2011.

7. Patents and China's Bt Rice

This chapter was coauthored with Lijun Liu. Skipping some technical details of this chapter will not affect one's reading of the rest of the chapter.

1. Clive James, "China Approves Biotech Rice and Maize in Landmark Decision," *Crop Biotech Update*, December 4, 2009. accessed January 10, 2013, http://www.isaaa.org/kc/cropbiotechupdate/article/default.asp?ID=5112.

2. Li Siwen et al., "Public Reception of and Their Attitudes and Behavior Toward GM Rice" (in Chinese), in *Proceedings of Interdisciplinary Workshop on Genetic Engineering Technology*, ed. *Journal of the Dialectics of Nature* and Nanjing Agricultural University (Nanjing, Jiangsu: April 15–16, 2012), 58–69.

3. Zhou Yijun et al., "Investigation on the Intellectual Property Rights About Transgenic Bt Rice" (in Chinese), in *Risk Assessment and Safety Regulation of Genetically Modified Organisms: Proceedings of the Third Workshop of the International Biosafety Forum*, ed. Xue Dayuan (Beijing: China Environmental Science Press, 2009), 133–169.

4. Greenpeace and Third World Network, *Chinese GM Rice Trapped in Foreign Patents* (in Chinese), accessed February 20, 2013, http://www.greenpeace.org/china/Global/china/_planet-2/report/2008/5/ge-patent-report.pdf.

5. Greenpeace and Third World Network, *Who Is the Real Owner of Chinese GM Rice?* (in Chinese), accessed February 20, 2013, http://www.greenpeace.org/china/Global/china/_planet-2/report/2009/2/3045095.pdf.

6. Jia Hepeng, "Chinese Green Light for GM Rice and Maize Prompts Outcry," *Nature Biotechnology* 28, no. 5 (May 2010): 390–391.

7. Announcement on the intellectual property rights issues of insect-resistant GM rice and phytase maize is available online at http://www.moa.gov.cn/ztzl/zjyqwgz/zswd/201007/t20100717_1601272.htm (accessed February 20, 2013).

8. SIPO, "The Argument of Foreign Companies' Control of Chinese GM Crop Patents Is Not Right" (in Chinese), *Legal Daily*, March 23, 2010.

9. Liu Xuxia and Li Jieyu, "The Intellectual Property Issues of the Commercialization of Transgenic Rice in China: A Response to the Question of 'Foreign Patent Trap'" (in Chinese), *Chinese Bulletin of Life Science* 23, no. 2 (February 2011): 223.

10. While acknowledging Li Hui and Christoph Then, respectively, in their research, the Greenpeace/Third World Network reports do not indicate who was commissioned to do the research.

11. Yanqing Wang and Sam Johnston, "The Status of GM Rice R&D in China," *Nature Biotechnology* 25, no. 7 (July 2007): 717–718.

12. Zhang Siguang, Shen Xiaobai, and Duan Yibing, "Public-Private Partnerships in Commercialization of Biotechnology: The Case of Transgenic Rice in China" (in Chinese), *Scientific Research Management* 27, no. 5 (2006): 71–76.

13. Jikun Huang et al., "Insect-Resistant GM Rice in Farmers' Fields: Assessing Productivity and Health Effects in China," *Science* 308 (April 29, 2005): 688–690.

14. Mao Chen, Anthony Shelton, and Gong-yin Ye, "Insect-Resistant Genetically Modified Rice in China: From Research to Commercialization," *Annual Review of Entomology* 56 (2011): 86.

15. Wang and Johnston, "The Status of GM Rice R&D in China."

16. Richard Stone, "China Plans $3.5 Billion GM Crops Initiative," *Science* 321 (September 5, 2008): 1279.

17. Liu and Li, "The Intellectual Property Issues of the Commercialization of Transgenic Rice in China."

18. As Zhang's Ph.D. student at that time, Tu conducted most of his work when he visited the Philippines-based International Rice Research Institute between 1995 and 1998; interview, Hangzhou, July 18, 2013.

19. Liu and Li, "The Intellectual Property Issues of the Industrialization of Transgenic Rice in China."

20. Readers may choose to skip following several paragraphs because of their technical nature.

21. Chen, Shelton, and Ye, "Insect-Resistant Genetically Modified Rice in China."

22. Jumin Tu et al., "Field Performance of Transgenic Elite Commercial Hybrid Rice Expressing *Bacillus thuringiensis* Delta-Endotoxin," *Nature Biotechnology* 18, no. 10 (2000): 1101–1104.

23. A. Depicker et al., "Nopaline Synthase: Transcript Mapping and DNA Sequence," *Journal of Molecular & Applied Genetics* 1, no. 6 (1982): 561–573.

24. European Parliament and Council, *Legal Protection of Biotechnological Inventions*, European Parliament and Council Directive 98/44 (Brussels: 1988), Article 9.

25. This international agreement aims at sharing the benefits arising from the utilization of genetic resources in a fair and equitable way, including by appropriate access to genetic resources and by appropriate transfer of relevant technologies, taking into account all rights over those resources and to technologies. It was adopted by the Conference of the Parties to the Convention on Biological Diversity at its tenth meeting, on October 29, 2010, in Nagoya, Japan, and became effective on October 12, 2014, almost four years after it was agreed. "The Nagoya Protocol on Access and Benefit-Sharing," Convention on Biological Diversity, accessed July 20, 2015, http://www.cbd.int/abs/.

26. Kong-Ming Wu et al., "Suppression of Cotton Bollworm in Multiple Crops in China in Areas with Bt Toxin–Containing Cotton," *Science* 321 (September 19, 2008): 1676–1678.

27. Ruifa Hu et al., "Reforming Intellectual Property Rights and the Bt Cotton Seed Industry in China: Who Benefits from Policy Reform?," *Research Policy* 38 (2009): 793–801; Carl Pray et al., "Impact of Bt Cotton in China," *World Development* 29, no. 5 (2001), 813–825.

28. Liu Jianqiang, "GM Rice for the 1.3 Billion Chinese: A Game of Safety and Interests" (in Chinese), *Southern Weekend*, December 9, 2004.

29. Dai Renmiao and Lu Xiaohui, "RMB24 Billion Investment in GMOs Might Be Wasted Because of Foreign Patent Monopoly" (in Chinese), *Twenty-First Business Herald*, April 19, 2013.

30. Interview with a Greenpeace China campaigner, Beijing, October 29, 2010.

8. China as a GMO Nation

1. Robert L. Paarlberg, *The Politics of Precaution: Genetically Modified Crops in Developing Countries* (Baltimore: Johns Hopkins University Press, 2001).

2. Daniel Lee Kleinman, "Conceptualizing the Politics of Science: A Response to Cambrosio, Limoges and Pronovost," *Social Studies of Science* 21, no. 4 (1991): 769–774.

3. Interview with a plant ecologist, Beijing, November 15, 2010.

4. See, for example, Yiyi Lu, "Environmental Civil Society and Governance in China," *International Journal of Environmental Studies* 64, no. 1 (2007): 59–69; and Andrew C. Mertha, *China's Water Warriors: Citizen Action and Policy Change* (Ithaca, NY: Cornell University Press, 2008).

5. On January 22, 2003, Green Earth Volunteers, an environmental NGO, featured Xue Dayuan of SEPA and Zhu Jiangang of Greenpeace at one of its salons for environmental journalists. See Guobin Yang and Craig Calhoun, "Media, Civil Society, and the Rise of a Green Public Sphere in China," *China Information* 21, no. 2 (2007): 219; and "Biosafety Management of GMOs: Current Situations and Trends" (in Chinese), Green Earth Volunteers, accessed October 20, 2014, http://www.chinagev.org/index.php/greenpro/huanjingjizhe/1508-zhuanjiyinshengwuguanli.

6. Dayuan Xue, *A Summary of Research on the Environmental Impacts of Bt Cotton in China* (Hong Kong: Greenpeace, 2002).

7. Massimiano Bucchi and Federico Neresini, "Why Are People Hostile to Bio-technologies?," *Science* 304 (June 18, 2004): 1749.

8. Matthew C. Nisbet and Bruce V. Lewenstein, "Biotechnology and the American Media: The Policy Process and the Elite Press, 1970 to 1999," *Science Communication* 23 (2002): 383.

9. Funing Zhong et al., "GM Food: A Nanjing Case Study of Chinese Consumers' Awareness and Potential Attitude," *AgBioForum* 5, no. 4 (2002): 136–144.

10. Lester Ross and Walter Hutchens, "Genetic Modification in Agriculture: The Impact of China's Regulations on Foreign Trade and Investment," *China Law & Practice* 15, no. 9 (2001): 30–37; Mary A. Marchang, Cheng Fang, and Baohui Song, "Issues on Adoption, Import Regulations, and Policies for Biotech Commodities in China with a Focus on Soybeans," *AgBioForum* 5, no. 4 (2002): 167–174.

11. Peng Liguo and Li Xudong, "Search for the Truth of Safety of Transgene in China: The Hidden Secret" (in Chinese), *Southern Weekend*, May 13, 2011.

12. Alasdair R. Young, "Political Transfer and 'Trading Up'? Transatlantic Trade in Genetically Modified Food and U.S. Politics," *World Politics* 55, no. 4 (2003): 457–484.

13. Tao Zhang and Shudong Zhou, "The Economic and Social Impact of GMOs in China," *China Perspectives*, no. 47 (June 2003): 50–57.

14. Niu Shuping, David Stanway, and Naveen Thukral, "China Rejects More U.S. Corn Due to GMO As State Sales Approach," *Reuters*, March 25, 2014, accessed September 25, 2014, http://www.reuters.com/article/2014/03/25/china-corn-usa-idUSL4N0MM0KY20140325; Jacob Bunge, "China's Limits on GMO Corn Drive Rift in U.S. Farm Industry," *Wall Street Journal*, April 11, 2014.

15. Hu Junhua, "China Approves Twelve GM Corn for Imports" (in Chinese), *China Business News*, July 4, 2014.

16. Tom Hancock, "China Approves Imports of New US-Developed GM Crops," *Financial Times*, June 15, 2017.

17. Chuin-Wei Yap, "It's China vs. China in Genetically Modified Food Fight," *China Real Time Report* (blog). *Wall Street Journal*, September 3, 2013, accessed August 18, 2014, http://blogs.wsj.com/chinarealtime/2013/09/03/its-china-vs-china-in-gmo-food-fight/.

18. Guy Cook, *Genetically Modified Language* (London: Routledge, 2005).

19. Jinjie Yang, Kaibin Xu, and Lulu Rodriguez, "The Rejection of Science Frames in the News Coverage of the Golden Rice Experiment in Hunan, China," *Health, Risk & Society* 16, no. 4 (2014): 339–354.

20. US-China Business Council, "USCBC Recommendations for Revisions to the 2014 Draft Catalogue Guiding Foreign Investment in Industry," December 3, 2014, accessed December 20, 2015, https://www.uschina.org/sites/default/files/USCBC%20Recommendations%202014%20Draft%20Catalogue%20Guiding%20Foreign%20Investment%20-%20English.pdf.

21. Interview with an agricultural policy analyst, Beijing, August 30, 2011; interview with a Monsanto staff member, Beijing: November 26, 2010.

22. Emily Waltz, "China's GM Rice First," *Nature Biotechnology* 28, no. 8 (2010): 8.

23. ESRC Global Environmental Change Programme, *The Politics of GM Food: Risk, Science & Public Trust*, Special Briefing No 5 (Swindon: ESRC, October 1999), 20.

24. Robert K. Merton, "The Normative Structure of Science," in *The Sociology of Science: Theoretical and Empirical Investigations*, ed. Norman W. Store (Chicago: University of Chicago Press, 1973 [1942]), 268–269.

25. John Ziman, "Research as a Career," in *The Research System in Transition*, ed. Susan E. Cozzens et al. (Dordrecht: Kluwer, 1990), 345–359; John Ziman, *Real Science: What It Is and What It Means* (Cambridge: Cambridge University Press, 2000).

26. Michael Gibbons et al., *The New Production of Knowledge: The Dynamics of Science and Research in Contemporary Societies* (London: Sage, 1994); Helga Nowotny, Peter Scott, and Michael Gibbons, *Re-Thinking Science: Knowledge and the Public in an Age of Uncertainty* (Cambridge: Polity Press, 2001).

27. Nowotny, Scott, and Gibbons, *Re-Thinking Science*, 47, 53.

28. Nowotny, Scott, and Gibbons, *Re-Thinking Science*, 96.

29. Roger Pielke Jr., *The Honest Broker: Making Sense of Science and Policy and Politics* (Cambridge: Cambridge University Press, 2007).

30. Thomas Kuhn, *The Structure of Scientific Revolution* (Chicago: University of Chicago Press, 1970).

31. Silvio O. Funtowicz and Jerome R. Ravetz, "A New Scientific Methodology for Global Environmental Issues," in *Ecological Economics: The Science and Management of Sustainability*, ed. Robert Costanza (New York: Columbia University Press, 1991), 138.

32. Nowotny, Scott, and Gibbons, *Re-Thinking Science*, 12.

33. Alison Henderson, C. Kay Weaver, and George Cheney, "Talking 'Facts': Identity and Rationality in Industry Perspectives on Genetic Modification," *Discourse Studies* 9, no. 1 (February 2007): 10.

34. David Toke and David Marsh, "Policy Networks and the GM Crops Issue: Assessing the Utility of a Dialectical Model of Policy Networks," *Public Administration* 81, no. 2 (2003): 229–251; Éric Montpetit, "A Policy Network Explanation of Biotechnology Policy Differences Between the United States and Canada," *Journal of Public Policy* 25, no. 3 (2005): 339–366.

35. ESRC Global Environmental Change Programme, *The Politics of GM Food*, 14.

36. Interviews with an agricultural policy analyst, Beijing, November 17, 2010, and August 30, 2011.

37. Erik Baark, "The Making of Science and Technology Policy in China," *International Journal of Technology Management* 21, no. 1/2 (2001): 1–21; Richard P. Suttmeier and Cong Cao, "China's Technical Community: Market Reform and the Changing Policy Cultures of Science," in *Chinese Intellectuals Between Market and State*, ed. Edward Gu and Merle Goldman (London: Routledge-Curzon, 2004), 138–157.

38. Anthony Saich, *Providing Public Goods in Transitional China* (New York: Palgrave Macmillan, 2008), 195.

39. Xia Shang, *Science Communication Behaviors of Greenpeace China on Genetically Modified Organism Issues* (in Chinese) (master's thesis, Peking University, Beijing, 2007); interview with an agricultural economist, Beijing: November 3, 2010.

40. Mertha, *China's Water Warriors*.

41. Nowotny, Scott, and Gibbons, *Re-Thinking Science*, 23–24.

42. See, for example, Sheila Jasanoff, *Designs on Nature: Science and Democracy in Europe and the United States* (Princeton, NJ: Princeton University Press, 2005).

43. For discussion on frame and political mobilization in social movements, see, for example, David A. Snow et al., "Frame Alignment Processes, Mobilization and Movement Participation," *American Sociological Review* 51, no. 4 (1986): 464–481; and Robert D. Benford and David A. Snow, "Framing Processes and Social Movements: An Overview and Assessment," *Annual Review of Sociology* 26 (2000): 611–639.

44. Sheila Jasanoff, "Product, Process or Programme: Three Cultures and the Regulation of Biotechnology," in *Resistance to New Technology, Nuclear Power, Information Technology and Biotechnology*, ed. Martin W. Bauer (Cambridge: Cambridge University Press, 1995), 311–331.

45. For a discussion of "institutional uncertainty" facing innovation in China, see Dan Breznitz and Michael Murphree, *Run of the Red Queen: Government, Innovation, Globalization, and Economic Growth in China* (New Haven, CT: Yale University Press, 2011).

46. Interview with a Monsanto staff member, Beijing, November 26, 2010.

47. Interview with a Monsanto staff member, Beijing, November 26, 2010.

48. Ma Aiping, "More than 90 Million Tons of Soybeans Imported Last Year to Ensure Chinese People Eat Well" (in Chinese), *Science and Technology Daily*, January 31, 2018.

49. Eduardo J. Trigo and Eugenio J. Cap, "Transforming Agriculture in Argentina: The Role of Genetically Modified (GM) Crops," in *Successful Agricultural Innovation in Emerging Economics: New Genetic Technologies for Global Food Production*, ed. David J. Bennett and Richard C. Jennings (Cambridge: Cambridge University Press, 2013), 140–141, 147.

50. ESRC Global Environmental Change Programme, *The Politics of GM Food*, 13.

51. Wen Jin, "Zhangye in Gansu Issued Document Banning GM Crop Growing" (in Chinese), *Jinghua Times*, November 1, 2013.

52. Xinhua News Agency, "Heilongjiang Will Forbid Planting GM Food Crops in Next Year (2017)" (in Chinese), December 26, 2016, accessed June 10, 2017, http://news.xinhuanet.com/politics/2016-12/16/c_1120134366.htm.

53. Susanna Hornig Priest, "Structuring Public Debate on Biotechnology," *Science Communication* 16, no. 2 (1994): 166–179.

54. Heidi Ledford, "Gao Caixia: Crop Engineer," *Nature* 534 (June 23 2016): 459.

55. Jacob Bunge and Lucy Craymer, "Scientists in China Race to Edit Crop Genes Sowing Unease in US," *Wall Street Journal*, May 6, 2018.

56. Fei Han et al., "Attitudes in China About Crops and Foods Developed by Biotechnology," *PLoS ONE* 10, no. 9 (September 29, 2015): e0139114; Chen Xuan and Chen Jie, "Trust, Risk Perception and Public Acceptance of Science and Technology: A Discussion on GM Rice in China" (in Chinese), *Science and Society* 6, no. 1 (2016): 93–109.

Bibliography

Achenbach, Joel. "107 Nobel Laureates Sign Letter Blasting Greenpeace Over GMOs." *Washington Post*, June 30, 2016.

An Baijie. "Court Rejects Lawsuit Over GM Rice: Lawyer." *Global Times*, March 26, 2010.

Anderson, Kym, and Shunli Yao. "China, GMOs, and World Trade in Agricultural and Textile Products." *Pacific Economic Review* 8, no. 2 (2003): 157–169.

Andsager, Julie L. "How Interest Groups Attempt to Shape Public Opinion with Competing News Frames." *Journalism & Mass Communication Quarterly* 77, no. 3 (2000): 572–592.

Associated Press (AP). "Shell Suppliers Give $60 Million to Taco Bell." June 8, 2001. AP News Archive. Accessed August 10, 2014. http://www.apnewsarchive .com/2001/Shell-Suppliers-Give-$60M-to-Taco-Bell/id-55902df5ddc5ae 7a80bb61bd1b5faa0a.

Baark, Erik. "The Making of Science and Technology Policy in China." *International Journal of Technology Management* 21, no. 1/2 (2001): 1–21.

Ball, Molly. "Want to Know If Your Food Is Genetically Modified?" *Atlantic*, May 14, 2014. Accessed July 29, 2014. http://www.theatlantic.com/features/archive/2014/05 /want-to-know-if-your-food-is-genetically-modified/370812/.

Bao Xiaodong, Li Yifan, and Wang Yue. "An Unexpected Disturbance Following a Joint Petition: Looking for 61 'Pro-GMO' Academicians" (in Chinese). *Southern Weekend*, December 5, 2013.

Bauer, Martin W. "Controversial Medical and Agri-Food Biotechnology: A Cultivation Analysis." *Public Understanding of Science* 22, no. 2 (2002): 93–111.

——. "The Evolution of Public Understanding of Science: Discourse and Comparative Evidence." *Science, Technology & Society* 14, no. 2 (2009): 221–240.

Beck, Ulrich. *Risk Society: Towards a New Modernity.* Trans. Mark Ritter. Los Angeles: Sage, 1992.

Benford, Robert D., and David A. Snow. "Framing Processes and Social Movements: An Overview and Assessment." *Annual Review of Sociology* 26 (2000): 611–639.

Berg, Paul. "Asilomar and Recombinant DNA." Stockholm: Nobel Foundation, August 26, 2004. Accessed July 29, 2014. http://www.nobelprize.org/nobel _prizes/chemistry/laureates/1980/berg-article.html.

Berg, Paul, and Janet E. Mertz. "Personal Reflections on the Origins and Emergence of Recombinant DNA Technology." *Genetics* 184, no. 1 (January 2010): 9–17.

Black, Robert E., Lindsay H. Allen, Zulfiqar A. Bhutta, Laura E. Caulfield, Mercedes de Onis, Majid Ezzati, Colin Mathers, and Juan Rivera, for the Maternal and Child Undernutrition Study Group. "Maternal and Child Undernutrition: Global and Regional Exposures and Health Consequences." *Lancet* 371, no. 9608 (January 19, 2008): 243–260.

Bloomberg News. "Bayer Settles with Farmers Over Modified Rice Seeds." *New York Times*, July 2, 2011.

Boland, Alana. "The Three Gorges Debate and Scientific Decision-Making in China." *China Information* 13, no. 1 (1998): 25–42.

Branigan, Tania. "China Shut Down Maoist Website Utopia." *Guardian*, April 6, 2012.

Breznitz, Dan, and Michael Murphree. *Run of the Red Queen: Government, Innovation, Globalization, and Economic Growth in China.* New Haven, CT: Yale University Press, 2011.

Brookes, Graham. "The Economic and Environmental Impact of First-Generation Biotech Crops." In *Successful Agricultural Innovation in Emerging Economics: New Genetic Technologies for Global Food Production*, ed. David J. Bennett and Richard C. Jennings, 61–81. Cambridge: Cambridge University Press, 2013.

Brooks, Harvey. "The Scientific Adviser." In *Scientists and National Policy-Making*, ed. Robert Gilpin and Christopher Wright, 76–77. New York: Columbia University Press, 1964.

Brossard, Dominique, and James Shanahan, "Do Citizens Want to Have Their Say? Media, Agricultural Biotechnology, and Authoritarian Views of Democratic Process in Science." *Mass Communication & Society* 6, no. 3 (2003): 291–312.

Bucchi, Massimiano, and Federico Neresini. "Why Are People Hostile to Biotechnologies?" *Science* 304 (June 18, 2004): 1749.

Bunge, Jacob. "China's Limits on GMO Corn Drive Rift in U.S. Farm Industry." *Wall Street Journal*, April 11, 2014.

Bureau of Rural and Social Development and China National Center for Biotechnology Development, under the Ministry of Science and Technology, the

People's Republic of China. *China's Biotechnology Development Report* 2002 (in Chinese). Beijing: China Agricultural Press, 2003.

Buss, Dale. "American Still Aren't Buying GMO-Free Gospel. Just Ask General Mills." *Forbes*, February 20, 2014.

Cantley, Mark F., and Drew L. Kershen. "Regulatory Systems and Agricultural Biotechnology." In *Successful Agricultural Innovation in Emerging Economies: New Genetic Technologies for Global Food Production*, ed. David J. Bennett and Richard C. Jennings, 267–282. Cambridge: Cambridge University Press, 2013.

Cao, Cong. *China's Scientific Elite*. London: RoutledgeCurzon, 2004.

——. "Chinese Science and the 'Nobel Prize Complex.'" *Minerva* 42, no. 2 (2004): 151–172.

——. "Strengthening China Through Science and Education: China's New Development Strategy Toward the Twenty-First Century." *Issues & Studies* 38, no. 3 (2002): 122–149.

Cao, Cong, Ning Li, Xia Li, and Li Liu. "Reforming China's S&T System." *Science* 341 (August 2, 2013): 460–462.

Cao, Cong, and Richard P. Suttmeier. "China's 'Brain Bank': Leadership and Elitism in Chinese Science and Engineering." *Asian Survey* 39, no. 3 (1999): 525–559.

——. "Challenges of S&T System Reform in China." *Science* 335 (March 10, 2017): 1019–1021.

Cao, Cong, Richard P. Suttmeier, and Denis Fred Simon. "China's 15-Year Science and Technology Plan." *Physics Today* 59, no. 12 (2006): 38–43.

Casassus, Barbara. "Study Linking GM Maize to Rat Tumours Is Retracted." *Nature*, November 28, 2013. Accessed August 31, 2014. http://www.nature.com/news/study-linking-gm-maize-to-rat-tumours-is-retracted-1.14268.

CCTV. *News Probe: Tracing GM Rice*. July 26, 2014. Accessed July 20, 2017. http://zhongyang13.com/cctv13/xinwendiaocha/20140727/10297.html.

Cendrowicz, Leo. "Is Europe Finally Ready for Genetically Modified Foods?" *Time*, March 9, 2010.

Center for Food Safety. "GE Food Labeling: States Take Action." Washington, DC: Center for Food Safety, June 2014. Accessed September 10, 2014. http://www.centerforfoodsafety.org/files/ge-state-labeling-fact-sheet-620141_28179.pdf.

——. "USDA Aims to Commercialize Genetically Engineered Sugar Beets, Ignores Documented Environmental Impacts." Washington, DC: Center for Food Safety, December 10, 2010. Accessed November 10, 2017. https://www.centerforfoodsafety.org/press-releases/817/usda-aims-to-commercialize-genetically-engineered-sugar-beets-ignores-documented-environmental-impacts.

Chang, Kenneth. "These Foods Aren't Genetically Modified but They Are 'Edited.'" *New York Times*, January 10, 2017.

Charles, Dan. "Golden Rice Study Violated Ethical Rules, Tufts Says." *National Public Radio*, September 17, 2013. Accessed October 30, 2014. http://www.npr.org/blogs/thesalt/2013/09/17/223382375/golden-rice-study-violated-ethical-rules-tufts-says.

Chen Hongxia, and Zhang Xu. "Biosafety Certificates Close to Renewal While Controversy Around Commercialization of Food Staple Continues" (in Chinese). *Twenty-First Century Business Herald*, August 19, 2014.

Chen, Jie. *Transnational Civil Society in China: Intrusion and Impact*. Cheltenham, UK: Edward Elgar, 2012.

Chen, Mao, Anthony Shelton, and Gong-yin Ye. "Insect-Resistant Genetically Modified Rice in China: From Research to Commercialization." *Annual Review of Entomology* 56 (2011): 81–101.

Chen Ming, Ye Biao, Shen Ying, Xie Xue, and Xiao Han. "Fang's Method: Fang Zhouzi and the Debate Rules Under His Influence" (in Chinese). *Southern Weekend*, June 22, 2012.

Chen, Nancy N. "Feeding the Nation: Chinese Biotechnology and Genetically Modified Food." In *Asian Biotech: Ethics and Communities of Fate*, ed. Aihwa Ong and Nancy N. Chen, 81–92. Durham, NC: Duke University Press, 2010.

Chen Weixiao. "China Made 'Landmark' GM Food Crop Approval." *SciDev.Net*, December 17, 2009. Accessed January 10, 2013. http://www.scidev.net/global/biotechnology/news/china-makes-landmark-gm-food-crop-approval.html.

Chen Xiwen. "Interpreting the Text of the No. 1 Central Document" (in Chinese). January 22, 2014. Accessed January 23, 2014. http://china.caixin.com/2014–01–22/100632019.html.

Chen Xuan and Chen Jie, "Trust, Risk Perception and Public Acceptance of Science and Technology: A Discussion on GM Rice in China" (in Chinese). *Science and Society* 6, no. 1 (2016): 93–109.

Chen Yu. "Experts Interpret Central Document No. 1: Push Forward Commercialization of GMO New Varieties" (in Chinese). *Science and Technology Daily*, February 3, 2010.

Chen, Zhu, Hong-Guang Wang, Zhao-Jun Wen, and Yihuang Wang. "Life Sciences and Biotechnology in China." *Philosophical Transactions of the Royal Society, B: Biological Sciences* 362 (2007): 947–957.

Choudhary, Bhagirath. "Development of Transgenic Bt Cotton Technology in India and China: A Policy Perspective." *Science and Public Policy* 28, no. 3 (2001): 219–229.

Collins, Steven W. *The Race to Commercialize Biotechnology: Molecules, Markets and the State in the United States and Japan*. London: Routledge, 2004.

Commission of the European Communities. *White Paper on Food Safety*. Brussels: European Union, January 12, 2000.

"Confronting the Bad News Bearers." *MIT Technology Review* 95, no. 7 (October 1992): 20–31.

Cong Yuhua. "RMB20 Billion to Break the GM Crop Deadlock" (in Chinese). *China Youth Daily*, September 24, 2008.

Conway, Gordon, and Gary Toenniessen. "Feeding the World in the Twenty-First Century." *Nature* 402 Supplement (December 2, 1999): C55–58.

Cook, Guy. *Genetically Modified Language*. London: Routledge, 2005.

Cressey, Daniel. "Debate Rages Over Herbicide's Cancer Risk." *Nature*, November 15, 2015. Accessed November 20, 2015. http://www.nature.com /news/debate-rages-over-herbicide-s-cancer-risk-1.18794.

——. "A New Breed." *Nature* 497 (May 2, 2013): 27–29.

——. "Widely Used Herbicide Linked to Cancer." *Nature*, March 24, 2015. Accessed May 1, 2015. http://www.nature.com/news/widely-used-herbicide-linked -to-cancer-1.17181.

Cui Yongyuan, "Cui Yongyuan's Proposals: Pursuing GMOs" (in Chinese). Accessed December 20, 2015. http://news.sina.com.cn/o/2014-03-03/053029605547 .shtml.

——. "Objection to the Composition of the Fifth NBCAGMO" (in Chinese). Weibo post, June 16, 2016. Accessed December 27, 2016. http://weibo.com /ttarticle/p/show?id=2309403987131688203084.

Cui Zheng, with intern reporters Yang Jie, Zhang Xia, and Will Spence. "Monsanto's Cautious China Strategy." *Caixin Online*, December 24, 2013. Accessed September 25, 2014. http://english.caixin.com/2013–12–24/100621301.html.

Dai Qing. *Yangtze! Yangtze!* Ed. Patricia Adams and John Thibodeau. Trans. Nancy Liu, Wu Mei, Sun Yougeng, and Zhang Xiaogang. London: Earthscan, 1994.

Dai Renmiao, and Lu Xiaohui. "RMB24 Billion Investment in GMOs Might Be Wasted Because of Foreign Patent Monopoly" (in Chinese). *Twenty-First Business Herald*, April 19, 2013.

D'Angelo, Paul. "News Framing as a Multiparadigmatic Research Program: A Response to Entman." *Journal of Communication* 52, no. 4 (December 2002), 870–888.

Dean, Donald H. "Biochemical Genetics of the Bacterial Insect-Control Agent Bacillus thuringiensis: Basic Principles and Prospects for Genetic Engineering." *Biotechnology & Genetic Engineering Reviews* 2, no. 1 (1984): 341–363.

Deng Li. "China Cautious Towards GM Rice Commercialization" (in Chinese). *Twenty-First Century Business Herald*, July 10, 2008.

Depicker, A., S. Stachel, P. Dhaese, P. Zambryski, and H. Goodman. "Nopaline Synthase: Transcript Mapping and DNA Sequence." *Journal of Molecular & Applied Genetics* 1, no. 6 (1982): 561–573.

"Doubts About GM Rice" (in Chinese). *Outlook Weekly*, February 8, 2010.

Du, Li, and Christen Rachul. "Chinese Newspaper Coverage of Genetically Modified Organisms." *BMC Public Health* 12 (2012): 326.

Dunwoody, Sharon. "The Science Writing Inner Club: A Communication Link Between Science and the Lay Public." In *Scientists and Journalists: Reporting Science as News*, ed. Sharon M. Friedman, Sharon Dunwoody, and Carol L. Rogers, 155–169. New York: Free Press, 1986.

Durant, John, Martin W. Bauer, and George Gaskell, ed. *Biotechnology in the Public Sphere: A European Sourcebook*. London: Science Museum, 1998.

Dyson, Freeman J. *The Sun, the Genome, and the Internet: Tools of Scientific Revolution*. Oxford: Oxford University Press, 1999.

Economy, Elizabeth. *The River Runs Black: The Environmental Challenge to China's Future*. 2nd ed. Ithaca, NY: Cornell University Press, 2010.

"Email of Zou Ping of MOA's Office of Agricultural Genetic Engineering Biosafety Administration to Lorena Luo of Greenpeace, July 24, 2008." In Wen Jiayun (the Third World Network), "MOA Called off Golden Rice Experiment in China in 2008" (in Chinese). Blog post, August 31, 2012. Accessed May 9, 2014. http://blog.sina.com.cn/s/blog_81b7cbe801014kt5.html.

Engdahl, F. William. "About F. William Engdahl—Biography." Accessed November 16, 2014. http://www.williamengdahl.com/about.php.

——. "Engdahl China Speaking Tour (6)" (in Chinese). Blog post, November 4, 2008. Accessed June 22, 2014. http://blog.sina.com.cn/s/blog_5d21454b0100b8ka.html.

——. *Seeds of Destruction: The Hidden Agenda of Genetic Manipulation*. Montreal: Global Research, 2007.

Enserink, Martin. "Golden Rice Not So Golden for Tufts." *Science*, September 18, 2013. Accessed May 10, 2014. http://news.sciencemag.org/asiapacific/2013/09/golden-rice-not-so-golden-tufts.

ESRC Global Environmental Change Programme. *The Politics of GM Food: Risk, Science & Public Trust*. Special Briefing No. 5. Swindon: ESRC, October 1999.

"EU Court Bans BASF's 'Amflora' GM Potato, Annuls Commission Approval." *DW*, December 13, 2013. Accessed November 10, 2014. http://www.dw.de/eu-court-bans-basfs-amflora-gm-potato-annuls-commission-approval/a-17293624.

"EU Science Advisor: 'Lots of Policies Are Not Based on Evidence.'" *Euractiv*, July 24, 2012. Accessed August 31, 2014. http://www.euractiv.com/innovation-enterprise/chief-scientifc-adviser-policy-p-interview-514074.

"EU Tightens Control of Chinese Rice Over GM Fears." *EU Business*, November 15, 2011. Accessed June 7, 2014. http://www.eubusiness.com/news-eu/china-biotech-food.diq.

European Commission. *A Decade of EU-Funded GMO Research (2001–2010)*. Luxembourg: Publications Office of the European Union, 2010.

——. *EC-Sponsored Research on Safety of Genetically Modified Organisms: A Review of Results*. Luxembourg: Publications Office of the European Union, 2001.

——. "Fact Sheet: Questions and Answers on EU's Policies on GMOs." April 25, 2015. Accessed June 1, 2016. http://europa.eu/rapid/press-release_MEMO-15-4778_en.htm.

———. "GMO Legislation." Accessed June 1, 2016. http://ec.europa.eu/food/plant /gmo/legislation/index_en.htm.

European Parliament and Council. *Legal Protection of Biotechnological Inventions*, European Parliament and Council Directive 98/44. Brussels: 1998.

Ewen, Stanley, and Árpád Pusztai. "Effect of Diets Containing Genetically Modified Potatoes Expressing *Galanthus nivalis* lectin on Rat Small Intestine." *Lancet* 354, no. 9187 (October 16, 1999): 1353–1354.

Fang Zhouzi. "This Shows the Power of Science, Remark in Beijing on the Occasion of the National GM Rice Tasting Session" (in Chinese). Blog post, August 17, 2014. Accessed August 18, 2014. http://fangzhouzi.baijia.baidu.com /article/26328.

Fell, Andy. "Nothing Ventured, Nothing Gained." *UC Davis Magazine* 21, no. 3 (Spring 2004). Accessed June 6, 2014. http://ucdavismagazine.ucdavis.edu/issues /sp04/feature_1.html.

Feltman, Rachel. "Why This Genetically Modified Mushroom Gets to Skip USDA Oversight." *Washington Post*, April 18, 2016.

Ferguson, Cat. "Rice Researcher in Ethics Scrape Threatens Journal with Lawsuit Over Coming Retraction." *Retraction Watch*, July 17, 2014. Accessed August 10, 2014. http://retractionwatch.com/2014/07/17/rice-researcher-in-ethics-scrape -threatens-journal-with-lawsuit-over-coming-retraction/.

Fladung, Matthias. "Cibus' Herbicide-Resistant Canola in European Limbo." *Nature Biotechnology* 34 (May 2016): 473–474.

Food and Agriculture Organization of the United Nations and World Health Organization. *Summary Report of Joint FAO/WHO Meeting on Pesticide Residues* (Geneva: FAO and WHO, May 16, 2016). Accessed July 30, 2017. http://www .who.int/foodsafety/jmprsummary2016.pdf?ua=1.

Fresco, Louise O. "The GMO Stalemate in Europe." *Science* 329 (February 22, 2013): 883.

Frewer, Lynn J., Susan Miles, and Roy Marsh. "The Media and Genetically Modified Foods: Evidence in Support of Social Amplification of Risk." *Risk Analysis* 22, no. 4 (2002): 701–711.

Friedman, Sharon M., Sharon Dunwoody, and Carol L. Rogers, eds. *Scientists and Journalists: Reporting Science as News*. New York: Free Press, 1986.

Fu Zhongwen. "The Progress on Biosafety Regulation of Agri-GMOs in China" (in Chinese). In *Risk Assessment and Risk Management of Genetically Modified Organisms: Proceedings of the Fifth Workshop of the International Biosafety Forum*, ed. Xue Dayuan, 39–45. Beijing: China Environmental Science Press, 2012.

Funtowicz, Silvio O., and Jerome R. Ravetz. "A New Scientific Methodology for Global Environmental Issues." In *Ecological Economics: The Science and Management of Sustainability*, ed. Robert Costanza, 137–152. New York: Columbia University Press, 1991.

Gamson, William L., and Andre Modigliani. "Media Discourse and Public Opinion on Nuclear Power: A Constructivist Approach." *American Journal of Sociology* 95, no. 1 (July 1989): 1–37.

Gans, Herbert J. *Deciding What's News: A Study of CBS Evening News, NBC Nightly News, Newsweek, and Time.* New York: Pantheon Book, 1979.

Garon, Francis, and Éric Montpetit. "Different Paths to the Same Result: Explaining Permissive Policies in the USA." In *The Politics of Biotechnology in North America and Europe: Policy Networks, Institutions, and Internationalization*, ed. Éric Montpetit, Christine Rothmayr, and Frédéric Varone, 61–82. Lanham, MD: Lexington Books, 2007.

Gaskell, George, and Martin W. Bauer. "Biotechnology in the Years of Controversy: A Social Scientific Perspective." In *Biotechnology 1996–1999: The Years of Controversy*, ed. George Gaskell and Martin W. Bauer, 3–14. London: Science Museum Press, 2001.

"Gene Editing in Legal Limbo in Europe." *Nature* 542 (February 22, 2017): 392.

Gibbons, Michael, Camille Limoges, Helga Nowotny, Simon Schwartzman, Peter Scott, and Martin Trow. *The New Production of Knowledge: The Dynamics of Science and Research in Contemporary Societies.* London: Sage, 1994.

Gilbert, Natasha. "A Hard Look at GM Crops." *Nature* 497 (May 2, 2013): 24–26.

Gillam, Carey. "House Passes Anti-GMO Labeling Law." *Reuters*, July 23, 2015. Accessed August 18, 2015. http://www.reuters.com/article/2015/07/23/us-usa-gmo-labeling-idUSKCN0PX17920150723.

——. "U.S. Food Makers to Seek Single Federal Standard for GMO Labeling." *Reuters*, January 13, 2014. Accessed August 10, 2014. http://www.reuters.com/article/2014/01/13/usa-gmo-labeling-idUSL2N0KN1NX20140113.

"GM Rice Cultivation: Why Should China Aim for the First" (in Chinese). *Outlook Weekly*, December 7, 2009.

"GM Rice Have Waited Five Years for Their Biosafety Certificates" (in Chinese). *People's Daily* Online live interview transcript, December 25, 2009. Accessed January 11, 2010. http://live.people.com.cn/note.php?id=662091225085841_ctdzb_008.

Goldman, Merle. *Sowing the Seeds of Democracy in China: Political Reform in the Deng Xiaoping Era.* Cambridge, MA: Harvard University Press, 1994.

Gonick, Cy, and Andrea Levy. "An Interview with Vandana Shiva." *Canadian Dimension*, 48, no. 4 (July/August 2014). Accessed August 31, 2014. http://canadiandimension.com/articles/6235/.

Graham, Ian. "Using Molecular Breeding to Improve Orphan Crops for Emerging Economies." In *Successful Agricultural Innovation in Emerging Economies: New Genetic Technologies for Global Food Production*, ed. David J. Bennett and Richard C. Jennings, 95–106. Cambridge: Cambridge University Press, 2013.

Green Earth Volunteers. "Biosafety Management of GMOs: Current Situations and Trends" (in Chinese). Accessed October 20, 2014. http://www.chinagev.org/index .php/greenpro/huanjingjizhe/1508-zhuanjiyinshengwuguanli.

Greenpeace. "Chinese Consumer Loses GE Food Labelling Case Against Nestlé but Vows to Continue." April 20, 2004. Accessed October 10, 2014. http://www.greenpeace.org/eastasia/news/stories/food-agriculture/2004 /chinese-consumer-loses-ge-food/.

Greenpeace and Third World Network. *Chinese GM Rice Trapped in Foreign Patents* (in Chinese). Accessed February 20, 2013. http://www.greenpeace.org/china /Global/china/_planet-2/report/2008/5/ge-patent-report.pdf.

——. *Who Is the Real Owner of Chinese GM Rice?* (in Chinese). Accessed February 20, 2013. http://www.greenpeace.org/china/Global/china/_planet-2/report/2009/2 /3045095.pdf.

Griffitts, Joel S., Stuart M. Haslam, Tinglu Yang, Stephan F. Garczynski, Barbara Mulloy, Howard Morris, Paul S. Cremer, Anne Dell, Michael J. Adang, and Raffi V. Aroian. "Glycolipids as Receptors for Bacillus thuringiensis Crystal Toxin." *Science* 307 (February 11, 2005): 922–925.

Gu Xiulin. *GMO War: Defending China's Food Security in the Twenty-First Century* (in Chinese). Beijing: Intellectual Property Publishing House, 2011.

——. Weibo post. Accessed August 16, 2016. http://blog.sina.com.cn/gxiulin.

Guo Yuhua. "Angel or Evil? Social and Cultural Perspective to GM Soybeans in China" (in Chinese). *Sociological Study*, no. 1 (2005): 84–112.

Gurevitch, Michael, and Mark R. Levy, eds. *Mass Communication Review Yearbook*, vol. 5. Thousand Oaks, CA: Sage, 1986.

Gutteling, Jan M., et al., "Media Coverage 1973–1996: Trends and Dynamics." In *Biotechnology: The Making of a Global Controversy*, ed. Martin W. Bauer and George Gaskell, 95–128. Cambridge: Cambridge University Press, 2002.

Guyton, Kathryn Z., Dana Loomis, Yann Grosse, Fatiha El Ghissassi, Lamia Benbrahim-Tallaa, Neela Guha, Chiara Scoccianti, Heidi Mattock, Kurt Straif, on behalf of the International Agency for Research on Cancer Monograph Work- ing Group, IARC. "Carcinogenicity of Tetrachlorvinphos, Parathion, Malathion, Diazinon, and Glyphosate." *Lancet Oncology* 16, no. 5 (2015): 490–491.

Hall, Stephen S. "Editing the Mushroom." *Scientific American* 314, no. 3 (2016): 56–63.

Han, Fei, Dingyang Zhou, Xiaoxia Liu, Jie Cheng, Qingwen Zhang, and Anthony M. Sheldon. "Attitudes in China About Crops and Foods Developed by Biotech- nology." *PLoS ONE* 10, no. 9 (September 29, 2015): e0139114.

Hancock, Tom. "China Approves Imports of New US-Developed GM Crops." *Financial Times*, June 15, 2017.

Hao Xin, and Richard Stone. "Q&A: China's Science Premier." *Science* 322 (October 17, 2008): 362–364.

He Jun. "China Made GM Rice Quietly Granted Biosafety Certificates" (in Chinese). *Daily Economic News*, December 3, 2012.

Heap, Brian. "Europe Should Rethink Its Stance on GM Crops." *Nature* 498 (June 27, 2013): 409.

Henderson, Alison, C. Kay Weaver, and George Cheney. "Talking 'Facts': Identity and Rationality in Industry Perspectives on Genetic Modification." *Discourse Studies* 9, no. 1 (2007): 9–41.

Henig, Robin Marantz. *A Monk and Two Peas: The Story of Gregor Mendel and the Discovery of Genetics*. London: Phoenix, 2001.

Herring, Ron. "China, Rice, and GMOs: Navigating the Global Rift on Genetic Engineering." *Asia-Pacific Journal* 7, no. 3 (January 2009). Accessed January 16, 2009. http://japanfocus.org/-Ron-Herring/3012.

Herring, Ronald J. "Stealth Seeds: Bioproperty, Biosafety, Biopolitics?" *Journal of Development Studies* 43, no. 1 (2007): 130–157.

Ho, Peter, and Eduard B. Vermeer. "Food Safety Concerns and Biotechnology: Consumers' Attitudes to Genetically Modified Products in Urban China." *AgBioForum* 7, no. 4 (2004): 158–175.

Ho, Peter, Jennifer H. Zhao, and Xue Dayuan. "Access and Control of Agrobiotechnology: Bt Cotton, Ecological Change and Risk in China" (in Chinese). In *Risk Assessment and Safety Regulation of Genetically Modified Organisms: Proceedings of the Third Workshop of the International Biosafety Forum*, ed. Xue Dayuan, Yoke Ling Chee, and Kun Xue, 171–186. Beijing: China Environmental Science Press, 2009.

Hornig, Susanna. "Reading Risk: Public Response to Print Media Accounts of Technological Risk." *Public Understanding of Science* 2, no. 2 (April 1993): 95–109.

Hu Junhua. "China Approves Twelve GM Corn for Imports" (in Chinese). *China Business News*, July 4, 2014.

Hu, Ruifa, Carl Pray, Jikun Huang, Scott Rozelle, Cunhui Fan, and Caiping Zhang. "Reforming Intellectual Property Rights and the Bt Cotton Seed Industry in China: Who Benefits from Policy Reform?" *Research Policy* 38, no. 5 (2009): 793–801.

Huang, Can, Celeste Amorim, Mark Spinoglio, Borges Gouveia, and Augusto Medina, "Organization, Program, and Structure: Analysis of the Chinese Innovation Policy Framework." *R&D Management* 34, no. 4 (2004): 367–387.

Huang, Jikun, Ruijian Chen, Jianwei Mi, Ruifa Hu, and Ellie Osir. "Farmers' Seed and Pest Control Management for Bt Cotton in China." In *Biotechnology and Agricultural Development: Transgenic Cotton, Rural Institutions and Resource-Poor Farmers*, ed. Robert Tripp, 105–134. London: Routledge, 2009.

Huang, Jikun, Ruifa Hu, Ruijian Chen, Jianwei Mi, Scott Rozelle, and Carl Pray. *Economic Impacts of Transgenic Biotechnology: Ten Years of Bt-Cotton in China* (in Chinese). Beijing: Science Press, 2010.

Huang, Jikun, Ruifa Hu, Carl Pray, Fangbin Qiao, and Scott Rozelle. "Biotechnology as an Alternative to Chemical Pesticides: A Case Study of Bt Cotton in China." *Agricultural Economics* 29, no. 1 (2003): 55–67.

Huang, Jikun, Ruifa Hu, Scott Rozelle, and Carl Pray. "Insect-Resistant GM Rice in Farmers' Fields: Assessing Productivity and Health Effects in China." *Science* 308 (April 29, 2005): 688–690.

Huang, Jikun, Ruifa Hu, Hans van Meijl, and Frank van Tongeren. "Biotechnology Boosts to Crop Productivity in China: Trade and Welfare Implications." *Journal of Development Economics* 75, no. 1 (2004): 27–54.

Huang, Jikun, Scott Rozelle, Carl Pray, and Qingfang Wang. "Plant Biotechnology in China." *Science* 295 (January 25, 2002): 674–677.

Huang, Jikun, and Qingfang Wang. "Biotechnology Policy and Regulation in China." IDS Working Paper 195. Brighton, UK: University of Sussex, Institute of Development Studies, 2003.

Hvistendahl, Mara. "China Aims to Sow a Revolution with GM Seed Takeover." *Science* 356 (April 7, 2017): 16–17.

"Ideological Debate in China: The Little Red Bookstore." *Economist*, February 5, 2009.

Jalonick, Mary Clare. "Congress OKs Bill Requiring First GMO Food Labels." *Washington Post*, July 14, 2016.

James, Clive. "China Approves Biotech Rice and Maize in Landmark Decision." *Crop Biotech Update*, December 4, 2009. Accessed January 10, 2013. http://www.isaaa.org/kc/cropbiotechupdate/article/default.asp?ID=5112.

——. *Global Status of Commercialized Biotech/GM Crops.* ISAAA Annual Briefs, 1996–2015. Accessed June 1, 2016. http://www.isaaa.org/resources/publications/briefs/default.asp.

——. *ISAAA Brief 51: 20th Anniversary of the Global Commercialization of Biotech Crops (1996 to 2015) and Biotech Crop Highlights in 2015.* Executive Summary. Accessed June 1, 2016. http://www.isaaa.org/resources/publications/briefs/51/executivesummary/default.asp.

——. *ISAAA, Brief 52: Global Status of Commercialized Biotech/GM Crops: 2016.* Executive Summary. Accessed June 10, 2017. http://www.isaaa.org/resources/publications/briefs/52/executivesummary/default.asp.

Jasanoff, Sheila. *Designs on Nature: Science and Democracy in Europe and the United States.* Princeton, NJ: Princeton University Press, 2005.

——. *The Fifth Branch: Science Advisers as Policymakers.* Cambridge, MA: Harvard University Press, 1998.

——. "Product, Process or Programme: Three Cultures and the Regulation of Biotechnology." In *Resistance to New Technology, Nuclear Power, Information Technology and Biotechnology*, ed. Martin W. Bauer, 311–331. Cambridge: Cambridge University Press, 1995.

Ji Tianqin, Zhou Zhimei, Tang Ailin, Xu Yuwen, Chang Ye, and Wang Hao. "Scholar Wang Lijun" (in Chinese). *Southern Cosmopolitan Weekly*, no. 48 (December 28, 2012).

Jia Hepeng. "Chinese Green Light for GM Rice and Maize Prompts Outcry." *Nature Biotechnology* 28, no. 5 (2010): 390–391.

Jia Hepeng. "GM Rice May Soon Be Commercialized." *China Daily*, January 27, 2005.

Jia Hepeng, Tan Yihong, and Fan Jingqun. *Handbook for Science Journalists Covering GMOs* (in Chinese). Beijing: Center for Science Media, 2012.

Jiang Jingsong et al. "An Appeal to Postpone the Commercialization of GM Staple Foods" (in Chinese). Blog post, March 11, 2011. Accessed November 10, 2014. http://blog.sciencenet.cn/home.php?mod=space&uid=224810&do=blog&id=301777.

Jin Wei. "XY335: Rapid Expansion of an American Seed" (in Chinese). *International Herald Leader*, September 21, 2010.

Jin Wei, and Yu Shengnan. "Rats Disappeared: An Investigation into Abnormal Animal Phenomenon in Shanxi and Jilin" (in Chinese). *International Herald Leader*, September 21, 2010.

John, Brian. "Golden Rice Feeding Trials Breach Medical Ethics Code." Blog post, no date. Accessed August 14, 2014. http://www.gmfreecymru.org/pivotal_papers/feedingrice.html.

Jones, Mark. "The Invention of the Recombinant DNA Technology." *LSF Magazine*, Summer 2013. https://medium.com/lsf-magazine/the-invention-of-recombinant-dna-technology-e040a8a1fa22.

Kahn, Joseph. "The Science and Politics of Super Rice." *New York Times*, October 22, 2002.

Karplus, Valerie J., and Xingwang Deng. *Agricultural Biotechnology in China: Origins and Prospects*. New York: Springer, 2008.

Kasperson, Roger E., Ortwin Renn, Paul Slovic, Halina S. Brown, Jacque Emel, Robert Goble, Jeanne X. Kasperson, and Samuel Ratick. "The Social Amplification of Risk: A Conceptual Framework." *Risk Analysis* 8, no. 2 (1988): 177–187.

Ke Bei. "The Ignorance and Prejudice toward GMOs" (in Chinese). *Southern Weekend*, July 22, 2011.

Keeley, James. "Biotech Developmental State? Investigating the Chinese Gene Revolution." IDS Working Paper 207. Brighton, UK: University of Sussex, Institute of Development Studies, 2003.

——. "Regulating Biotechnology in China: The Politics of Biosafety." IDS Working Paper 208. Brighton, UK: University of Sussex, Institute of Development Studies, 2003.

"Key Figures in the Development of GM Staple Foods Are Implicitly Connected to the Rockefeller Foundation" (in Chinese). *Business Week*, March 15, 2010, 15.

Kleinman, Daniel Lee. "Conceptualizing the Politics of Science: A Response to Cambrosio, Limoges and Pronovost." *Social Studies of Science* 21, no. 4 (1991): 769–774.

Kleinman, Daniel Lee, Abby J. Kinchy, and Robyn Autry. "Local Variation or Global Convergence in Agricultural Biotechnology Policy? A Comparative Analysis." *Science and Public Policy* 35, no. 5 (2009): 361–371.

Kloor, Keith. "The GMO-Suicide Myth." *Issues in Science and Technology* 30, no. 2 (Winter 2014). Accessed October 20, 2014. http://issues.org/30–2/keith/.

Kopicki, Allison. "Strong Support for Labeling Modified Foods." *New York Times*, July 23, 2013.

Kuhn, Thomas. *The Structure of Scientific Revolution*. Chicago: University of Chicago Press, 1970.

LaFraniere, Sharon. "Fighting Trend, China Is Luring Scientists Home." *New York Times*, January 6, 2012.

Leask, Julie, Claire Hooker, and Catherine King. "Media Coverage of Health Issues and How to Work More Effectively with Journalists: A Qualitative Study." *BMC Public Health* 10 (2010): 535.

Ledford, Heidi. "CRISPR, the Disruptor." *Nature* 522 (June 4, 2015): 20–24.

Levick, Richard. "Are GMO-Free Cheerios the First Dominos?" *Forbes*, January 9, 2014.

——. "Colorado and Oregon: The GMO Conversation Shifts from Risks to Benefits." *Forbes*, November 5, 2014.

Lewis, Tanya. "Chipotle and GMO Debate." *Washington Post*, May 4, 2015.

Li Min. *A Case Study of the Coverage of GM Foods and Crops in People's Daily from the International Perspective*. Master's thesis, Tsinghua University, Beijing, 2007.

Li Siwen, Zhang Meng, Ge Xin, and Wen Tao. "Public Reception of and Their Attitudes and Behavior Toward GM Rice" (in Chinese). In *Proceedings of Interdisciplinary Workshop on Genetic Engineering Technology*, ed. *Journal of the Dialectics of Nature* and Nanjing Agricultural University, 58–69. Nanjing, Jiangsu: April 15–16, 2012.

Li Tie. "Analysis of Chinese-Style Fallacy and Rumors about GMOs" (in Chinese). *Southern Weekend*, August 19, 2011.

Li Yanjie. "Greenpeace: Thirty Percent of Rice Samples in Wuhan Contains GM Ingredients" (in Chinese). *China Business Journal*, May 17, 2014.

Li, Yunhe, Yufa Peng, Eric M. Hallerman, and Kongming Wu. "Biosafety Management and Commercial Use of Genetically Modified Crops in China." *Plant Cell Report* 33, no. 4 (2014): 565–573.

Liang Wei. "The Hengyang Golden Rice Story" (in Chinese). *Time Weekly*, December 13, 2012.

——. "300 Netizens Tried Golden Rice in Wuhan" (in Chinese). *Southern Metropolitan News*, October 20, 2013.

Lieberthal, Kenneth, and Michel Oksenberg. *Policy Making in China: Leaders, Structures, and Processes*. Princeton, NJ: Princeton University Press, 1988.

Lim, Xiaozhi. "European Scientists Call for GMO Legislation Revamp to Prevent Stifling Genome Editing." Genetic Literacy Project, July 25, 2014. Accessed June 1, 2016. https://www.geneticliteracyproject.org/2014/07/25/european -scientists-call-for-gmo-legislation-revamp-to-prevent-stifling-genome -editing/.

Lin Gu. "Greenpeace Landed in China Peacefully" (in Chinese). *Oriental Outlook Weekly*, no. 4 (December 11, 2003): 26–27.

Liu, Feng-chao, Denis Fred Simon, Yu-tao Sun, and Cong Cao. "China's Innovation Policies: Evolution, Institutional Structure, and Trajectory." *Research Policy* 40, no. 7 (2011): 917–931.

Liu Jianqiang. "GM Rice for the 1.3 Billion Chinese: A Game of Safety and Interests" (in Chinese). *Southern Weekend*, December 9, 2004.

——. "GM Rice Surprisingly Appears on the Market" (in Chinese). *Southern Weekend*, April 4, 2005.

Liu Li, *Theories and Practice on the Formulation of Basic Research-Related Policy* (in Chinese). Beijing: Tsinghua University Press, 2007.

Liu Xuxia, and Li Jieyu. "The Intellectual Property Issues of the Commercialization of Transgenic Rice in China: A Response to the Question of 'Foreign Patent Trap'" (in Chinese). *Chinese Bulletin of Life Science* 23, no. 2 (February 2011): 221–225.

Liu Yi. "The Vice Minister Supporting GMOs Accused to Be Employed by an American Company" (in Chinese). *Beijing Youth Daily*, October 31, 2013, A8.

Liu Zhiwei, and Fan Jingqun. "Safety of GM Rice Verified from Different Aspects" (in Chinese). *Science and Technology Daily*, January 7, 2010.

Losey, John E., Linda S. Rayor, and Maureen E. Carter. "Transgenic Pollen Harms Monarch Larvae." *Nature* 399 (May 20, 1999): 214.

Lu, Yiyi. "Environmental Civil Society and Governance in China." *International Journal of Environmental Studies* 64, no. 1 (2007): 59–69.

Lü Yongyan. "Basic Research and Molecular Breeding in Central Document No. 1" (in Chinese). *Guangming Net*, January 22, 2014, accessed January 23, 2014. http://tech.gmw.cn/2014–01/22/content_10196782.htm.

Lü Yongyan. "Netizens Reveal That Wang Dafang Is a Spy of the GMO Group" (in Chinese). Blog post, March 12, 2012, accessed June 22, 2014. http://blog.sina.com.cn/s/blog_51508bd70102duyp.html.

Lynas, Mark. "GM Won't Yield a Harvest for the World." *Guardian*, June 19, 2008.

——. "Lecture to Oxford Farming Conference." January 3, 2013. Accessed June 1, 2016. http://www.marklynas.org/2013/01/lecture-to-oxford-farming-conference-3 -january-2013/.

Ma Aiping, "More than 90 Million Tons of Soybeans Imported Last Year to Ensure Chinese People Eat Well" (in Chinese). *Science and Technology Daily*, January 31, 2018.

Ma, Nan, and Jiancheng Guan. "An Exploratory Study on Collaboration Profiles of Chinese Publications in Molecular Biology." *Scientometrics* 65, no. 3 (2005): 343–355.

Macilwain, Colin. "Against the Grain." *Nature* 422 (March 13, 2003): 111–112.

Mao Lei, and Liao Wengen. "Wu Baoguo Chairs the NPC Standing Committee's Sixteen Seminar" (in Chinese). *People's Daily*, June 26, 2010.

Mao Yufei. "Demystifying the Secret of Ruling Hidden in the Red-Head Official Documents" (in Chinese). *Global People*, August 2014. Accessed August 15, 2014. http://news.sina.com.cn/c/sd/2014-08-15/120730690491.shtml.

Marchang, Mary A., Cheng Fang, and Baohui Song. "Issues on Adoption, Import Regulations, and Policies for Biotech Commodities in China with a Focus on Soybeans." *AgBioForum* 5, no. 4 (2002): 167–174.

Marks, Leonie A., Nicholas Kalaitzandonakes, Kevin Allison, and Ludmila Zakharova. "Media Coverage of Agrobiotechnology: Did the Butterfly Have an Effect?" *Journal of Agribusiness* 21, no. 1 (Spring 2003): 1–20.

Marks, Leonie A., Nicholas Kalaitzandonakes, Lee Wilkins, and Ludmila Zakharova. "Mass Media Framing of Biotechnology News." *Public Understanding of Science* 16, no. 2 (2007): 183–203.

Masip, Gemma, Maite Sabalza, Eduard Pérez-Massot, Raviraj Banakar, David Cebrian, Richard M. Twyman, Teresa Capell, Ramon Albajes, and Paul Christou1. "Paradoxical EU Agricultural Policies on Genetically Engineered Crops." *Trends in Plant Science* 18, no. 6 (June 2013): 312–324.

McBeath, Jerry, and Jenifer Huang McBeath. "The Socio-Political Environment of Biosafety Regulation in China: The Case of GMOs." In *Risk Assessment and Regulation of Genetically Modified Organisms: Proceeding of the Fourth Workshop of the International Biosafety Forum*, ed. Xue Dayuan, 104–134. Beijing: China Environmental Science Press, 2012.

McCook, Alison. "Golden Rice Paper Pulled After Judge Rules for Journal." *Retraction Watch*, July 30, 2015. Accessed August 30, 2015. http://retractionwatch .com/2015/07/30/golden-rice-paper-pulled-after-judge-rules-for-journal/.

Meng Dengke. "GM Rice Allowed to Plant: Bashful Policy Released Quietly" (in Chinese). *Southern Weekend*, December 10, 2009.

Mertha, Andrew C. *China's Water Warriors: Citizen Action and Policy Change*. Ithaca, NY: Cornell University Press, 2008.

Merton, Robert K. "The Normative Structure of Science." In *The Sociology of Science: Theoretical and Empirical Investigations*, ed. Norman W. Store, 267–278. Chicago: University of Chicago Press, 1973 [1942].

Miller, Neal E. "The Scientist's Responsibility for Public Information: A Guide to Effective Communication with the Media." In *Scientists and Journalists: Reporting Science as News*, ed. Sharon M. Freidman, Sharon Dunwoody, and Carol L. Rogers, 239–253. New York: Free Press, 1986.

Mills, C. Wright. *The Power Elite*. Oxford: Oxford University Press, 1956.

Ministry of Science and Technology. "China's Plant GMO Research and Commercialization Special Program Achieve Significant Progress" (in Chinese). April 16, 2007. Accessed June 21, 2014. http://www.most.gov.cn/ncs/ncsgzdt/200704 /t20070416_53742.htm.

Mitchell, Tom, Christian Shepherd, and Ralph Atkins. "Syngenta Offers ChemChina Seeds of GM Growth." *Financial Times*, February 3, 2016.

"MOA: The National GMO Biosafety Committee Established" (in Chinese). Accessed June 15, 2014. http://www.china.com.cn/chinese/PI-c/170495.htm.

Montenegro, Maywa. "CRISPR Is Coming to Agriculture—with Big Implications for Food, Farmers, Consumers and Nature." *Ensia*, January 28, 2016. Accessed June 1, 2016. http://ensia.com/voices/crispr-is-coming-to-agriculture-with-big -implications-for-food-farmers-consumers-and-nature/.

Montpetit, Éric. "A Policy Network Explanation of Biotechnology Policy Differences Between the United States and Canada." *Journal of Public Policy* 25, no. 3 (2005): 339–366.

"Nagoya Protocol on Access and Benefit-Sharing." Convention on Biological Diversity. Accessed July 20, 2015. http://www.cbd.int/abs/.

Nakanishi, Nao. "China Seen a Crouching Dragon in Biotechnology." *Reuters*, December 20, 2002.

Nasto, Barbara. "Enter the Dragon: China's Biopharmaceutical Clusters." *Biotech in China, Nature Biotechnology*, Supplement (December 2012): B5.

National Academies of Sciences, Engineering, and Medicine. *Genetically Engineered Crops: Past Experience and Future Prospects*. Washington, DC: National Academies Press, 2016.

National Bureau of Statistics of China. *China Statistical Yearbook*. Beijing: China Statistics Press. Various years.

National Research Council, Committee on Genetically Modified Pest-Protected Plants. *Genetically Modified Pest-Protected Plants: Science and Regulation*. Washington, DC: National Academy Press, 2000.

Nelkin, Dorothy. "The Culture of Science Journalism." *Society* 24, no. 6 (September/ October 1987): 17–25.

——. "Selling Science." *Physics Today* 43, no. 11 (1990): 41–46.

Newell, Peter. "Domesticating Global Policy on GMOs: Comparing India and China." IDS Working Paper 206. Brighton, UK: University of Sussex, Institute for Development Studies, 2003.

Nisbet, Matthew C., Dominique Brossard, and Adrianne Kroepsch. "Framing Science: The Stem Cell Controversy in an Age of Press/Politics." *Press/Politics* 8, no. 2 (2003): 36–70.

Nisbet, Matthew C., and Bruce V. Lewenstein. "Biotechnology and the American Media: The Policy Process and the Elite Press, 1970 to 1999." *Science Communication* 23, no. 4 (June 2002): 359–391.

Niu Shuping, David Stanway, and Naveen Thukral. "China Rejects More U.S. Corn Due to GMO As State Sales Approach." *Reuters*, March 25, 2014. Accessed September 25, 2014. http://www.reuters.com/article/2014/03/25 /china-corn-usa-idUSL4N0MM0KY20140325.

Nobel Foundation. "The Nobel Prize in Physiology or Medicine 1962." Accessed July 28, 2014. http://www.nobelprize.org/nobel_prizes/medicine /laureates/1962/.

Norris, Pippa. *A Virtuous Circle: Political Communication in Postindustrial Societies.* Cambridge: Cambridge University Press, 2000.

Nowotny, Helga, Peter Scott, and Michael Gibbons. *Re-Thinking Science: Knowledge and the Public in an Age of Uncertainty.* Cambridge: Polity Press, 2001.

Offenheiser, Ray, and Kimberly Pfeifer. "Forward." In *Biotechnology and Agricultural Development: Transgenic Cotton, Rural Institutions and Resource-Poor Farmers,* ed. Robert Tripp, xxi–xxiv. New York: Routledge, 2009.

Office of Agricultural Genetic Engineering Biosafety Administration Under the Ministry of Agriculture, Institute of Biotechnology Under the Chinese Academy of Agricultural Sciences, and China Society for Agricultural Biotechnology, comps. *Thirty Years of Transgenic Technology* (in Chinese). Beijing: China Agricultural Science and Technology Press, 2012.

Oosterveer, Peter. *Global Governance of Food Production and Consumption: Issues and Challenges.* Cheltenham, UK: Edward Elgar, 2007.

Organisation for Economic Co-operation and Development (OECD). *Recombinant DNA Safety Considerations: Safety Considerations for Industrial, Agricultural and Environmental Applications of Organisms Derived by Recombinant DNA Techniques.* Paris: OECD, 1986.

Paarlberg, Robert L. *The Politics of Precaution: Genetically Modified Crops in Developing Countries.* Baltimore: Johns Hopkins University Press, 2001.

Parker, Laura. "The GMO Labeling Battle Is Heating Up—Here's Why." *National Geographic,* January 12, 2014. Accessed August 10, 2014. http://news.national geographic.com/news/2014/01/140111-genetically-modified-organisms-gmo -food-label-cheerios-nutrition-science/.

Pechlaner, Gabriela. *Corporate Crops: Biotechnology, Agriculture, and the Struggle for Control.* Austin: University of Texas Press, 2012.

Peng Guangqian. "Expert's Eight Questions on the Transgenization of Staple Foods: Why Should China Introduce the Technology Blindly" (in Chinese). *Global Times,* August 21, 2013.

Peng Kefeng and Wang Shan. "Greenpeace's Rice Theft at Night" (in Chinese). *China Science Daily,* May 9, 2014.

Peng Liguo and Li Xudong. "Search for the Truth of Safety of Transgene in China: The Hidden Secret" (in Chinese). *Southern Weekend,* May 13, 2011.

Phillips, Tom. "China Passes Law Imposing Security Controls on Foreign NGOs." *Guardian,* April 28, 2016.

Picker, Leslie, Danny Hakim, and Michael J. de la Merced. "Bayer Deal for Monsanto Follows Agribusiness Trend, Raising Worries for Farmers." *New York Times*, September 14, 2016.

Pielke, Roger, Jr. *The Honest Broker: Making Sense of Science and Policy and Politics.* Cambridge: Cambridge University Press, 2007.

Piñeyro-Nelson, A., J. Van Heerwaarden, H. R. Perales, J. A. Serratos-Hernández, A. Rangel, M. B. Hufford, P. Gepts, A. Garay-Arroyo, R. Rivera-Bustamante, and E. R. Álvarez-Buylla. "Transgenes in Mexican Maize: Molecular Evidence and Methodological Considerations for GMO Detection in Landrace Populations." *Molecular Ecology* 18, no. 4 (February 2009): 750–761.

Pinholster, Ginger. "AAAS Board of Directors: Legally Mandating GM Food Labels Could 'Mislead and Falsely Alarm Consumers.' " American Association for the Advancement of Science (AAAS), October 25, 2012. Accessed April 20, 2014. http://www.aaas.org/news/aaas-board-directors-legally-mandating-gm-food -labels-could-%E2%80%9Cmislead-and-falsely-alarm.

Pinstrup-Andersen, Per, and Ebbe Schioler. *Seeds of Contention: World Hunger and the Global Controversy Over GM Crops.* Baltimore: Johns Hopkins University Press, 2000.

Poppy, Guy. "Why Europe Needs a Chief Science Advisor." *Conversation*, November 17, 2014. Accessed November 23, 2014. http://theconversation.com /why-europe-needs-a-chief-scientific-advisor-34313.

Pray, Carl E. "Public and Private Collaboration on Plant Biotechnology in China." *AgBioForum* 2, no. 1 (1999): 48–53.

Pray, Carl E., Danmeng Ma, Jikun Huang, and Fangbin Qiao. "Impact of Bt Cotton in China." *World Development* 29, no. 5 (2001): 813–825.

Pray, Carl E., Latha Nagarajan, Jikun Huang, Ruifa Hu, and Bharat Ramaswami. "The Impact of Bt Cotton and the Potential Impact of Biotechnology on Other Crops in China and India." *Frontiers of Economics and Globalization* 10 (2011): 83–114.

Pray, Carl E., Bharat Ramaswami, Jikun Huang, Prajakta Bengali, and Huazhu Zhang. "Costs and Enforcement of Biosafety Regulations in India and China." *Journal of Technology and Globalisation* 2, no. 1/2 (2004): 137–157.

Priest, Susanna Hornig. "Structuring Public Debate on Biotechnology." *Science Communication* 16, no. 2 (1994): 166–179.

"Project NCT00680212: Vitamin A Equivalence of Plant Carotenoids in Children." Accessed May 9, 2014. http://clinicaltrials.gov/ct2/show/record /NCT00680212.

"Q&A: Manmohan Singh: India's Scholar–Prime Minister Aims for Inclusive Development." *Science* 335 (February 24, 2012): 907–908.

Qiu, Huangguang, Jikun Huang, Carl Pray, and Scott Rozelle. "Consumers' Trust in Government and Their Attitudes Towards Genetically Modified Food: Empirical

Evidence from China." *Journal of Chinese Economic and Business Studies* 10, no. 1 (2012): 67–87.

Qiu, Jane. "China Sacks Officials Over Golden Rice Controversy." *Nature*, December 10, 2012. Accessed January 10, 2013. http://www.nature.com/news /china-sacks-officials-over-golden-rice-controversy-1.11998.

——. "Is China Ready for GM Rice?" *Nature* 455 (October 16, 2008): 850–852.

Qu Lulu. *Reporting Risks on GMOs: An Analysis of the 15-Year News of Genetic Risks in Science and Technology Daily* (in Chinese). Master's thesis, Graduate School of the Chinese Academy of Sciences, Beijing, 2009.

Quist, David, and Ignacio H. Chapela. "Transgenic DNA Introgressed Into Traditional Maize Landraces in Oaxaca, Mexico." *Nature* 414 (November 29, 2001): 541–543.

Raab, Carolyn, and Deana Grobe. "Labeling Genetically Engineered Food: The Consumer's Right to Know?" *AgBioForum* 6, no. 4 (2003): 155–161.

Rabesandratana, Tania. "E.U. Moves Closer to Enabling National Bans on GM Crops." *Science*, November 12, 2014. Accessed November 23, 2014. http://news.sciencemag. org/environment/2014/11/e-u-moves-closer-enabling-national-bans-gm-crops.

Rao, Chavali Kameswara. "Genetically Engineered Crops Would Ensure Food Security in India." In *Successful Agricultural Innovation in Emerging Economics: New Genetic Technologies for Global Food Production*, ed. David J. Bennett and Richard C. Jennings, 167–183. Cambridge: Cambridge University Press, 2013.

Rao Yi. "Development and Challenges of Chinese Science: Publications in International Life Sciences Journals" (in Chinese). *Twenty-First Century*, no. 1 (February 2002): 83–93.

Rao Yi, Zhang Daqing, and Li Runhong. *Tu Youyou and Artemisinin* (in Chinese). Beijing: China Science and Technology Publisher, 2015.

Robin, Marie-Monique. *The World According to Monsanto: Pollution, Corruption, and the Control of Our Food Supply*. Trans. George Holoch. New York: New Press, 2010.

Rosenthal, Elisabeth. "Under Pressure, Chinese Newspaper Pulls Exposé on a Charity." *New York Times*, March 24, 2002.

Ross, Lester, and Walter Hutchens. "Genetic Modification in Agriculture: The Impact of China's Regulations on Foreign Trade and Investment." *China Law & Practice* 15, no. 9 (2001): 30–37.

Rotman, David. "Why We Will Need Genetically Modified Foods." *MIT Technology Review* 117, no. 1 (January/February 2014): 28–37.

Royal Society Expert Reviewers of Pusztai Rat Experiments at the Rowett Institute from 1998–1999. "Expert Reviewers 1–6 Sent with Mail to A. Pusztai on May 10, 1999 and Revised Expert Review Sent to A. Pusztai on May 12, 1999." Accessed November 26, 2014. http://www.ask-force.org/web/Pusztai/Royal -Society-Experts-1-7-19990510.pdf.

Ru Peng. *Science and Technology Experts and Decision Making in Science and Technology: The Influence of Science and Technology in the Decision Making of the 863 Program* (in Chinese). Beijing: Tsinghua University Press, 2012.

Saich, Anthony. *Providing Public Goods in Transitional China*. New York: Palgrave Macmillan, 2008.

Sawahel, Wadgy. "Iranian Scientists Produce Country's First GM Rice." *SciDev.Net*, February 18, 2005. Accessed October 10, 2014. http://www.scidev.net/global /gm/news/iranian-scientists-produce-countrys-first-gm-rice.html.

Schneider, Laurence. *Biology and Revolution in Twentieth-Century China*. Lanham, MD: Rowan & Littlefield, 2003.

Séralini, Gilles-Éric, Dominique Cellier, and Joël Spiroux de Vendômois. "New Analysis of a Rat Feeding Study with a Genetically Modified Maize Reveals Signs of Hepatorenal Toxicity." *Archives of Environmental Contamination and Toxicology* 52, no. 4 (2007): 596–602.

Séralini, Gilles-Éric, Emilie Clair, Robin Mesnage, Steeve Gress, Nicolas Defarge, Manuela Malatesta, Didier Hennequin, and Joël Spiroux de Vendômois. "Long Term Toxicity of a Roundup Herbicide and a Roundup-Tolerant Genetically Modified Maize." *Food and Chemical Toxicology* 50, no. 11 (2012): 4221–4231.

Shanahan, James. "Television and Authoritarianism: Exploring the Concept of Mainstreaming." *Political Communication* 15, no. 4 (1998): 483–495.

——. "Television Viewing and Adolescent Authoritarianism." *Journal of Adolescence* 18, no. 3 (June 1995): 271–288.

Shiva, Vandana. "From Seeds of Suicide to Seeds of Hope: Why Are Indian Farmers Committing Suicide and How Can We Stop This Tragedy?" *HuffPost*. May 29, 2009, updated May 25, 2011. Accessed August 31, 2014. http://www.huffington-post.com/vandana-shiva/from-seeds-of-suicide-to_b_192419.html.

Showalter, Ann M., Shannon Heuberger, Bruce E. Tabashnik, and Yves Carrière. "Development, Agronomic Performance and Sustainability of Transgenic Cotton for Insect Control." In *Biotechnology and Agricultural Development: Transgenic Cotton, Rural Institutions and Resource-Poor Farmers*, ed. Robert Tripp, 49–71. London: Routledge, 2009.

Simon, Denis Fred, and Cong Cao. *China's Emerging Technological Edge: Assessing the Role of High-End Talent*. Cambridge: Cambridge University Press, 2009.

Simons, Craig. "Of Rice and Men." *Newsweek*, December 20, 2004. Accessed June 15, 2014. http://www.newsweek.com/rice-and-men-123141.

Singh, Santosh K. *India: Agricultural Biotechnology Annual, 2011*. GAIN Report Number IN1187. Washington, DC: USDA Foreign Agricultural Service, Global Agricultural Information Network [GAIN], September 15, 2011.

——. *India: Agricultural Biotechnology Annual, 2012*. GAIN Report Number IN2098. Washington, DC: USDA Foreign Agricultural Service, Global Agricultural Information Network [GAIN], July 17, 2012.

———. *India: Agricultural Biotechnology Annual, 2013*. GAIN Report Number IN3083. Washington, DC: USDA Foreign Agricultural Service, Global Agricultural Information Network [GAIN], July 15, 2013.

———. *India: Agricultural Biotechnology Annual, 2014*. GAIN Report Number IN4059. Washington, DC: USDA Foreign Agricultural Service, Global Agricultural Information Network [GAIN], July 11, 2014.

———. *India: Agricultural Biotechnology Annual, 2015*. GAIN Report Number IN5088. Washington, DC: USDA Foreign Agricultural Service, Global Agricultural Information Network [GAIN], July 10, 2015.

———. *India: Agricultural Biotechnology Annual, 2016*. GAIN Report Number IN6157. Washington, DC: USDA Foreign Agricultural Service, Global Agricultural Information Network [GAIN], December 14, 2016.

Slovic, Paul. "Trust, Emotions, Sex, Politics, and Science: Surveying the Risk-Assessment Battlefield." In *The Perception of Risk*, ed. Paul Slovic, 390–412. Sterling, VA: Earthscan, 1997.

Smith, Jeffrey M. *Genetic Roulette: The Documented Health Risks of Genetically Engineered Foods*. 4th ed. Fairfield, IA: Yes! Books, 2007.

———. *Seeds of Deception: Exposing Industry and Government Lies About the Safety of the Genetically Engineered Foods You're Eating*. 3d ed. Fairfield, IA: Yes! Books, 2003.

Snow, David A., Burke Rochford, Steven Worden, and Robert Benford. "Frame Alignment Processes, Mobilization and Movement Participation." *American Sociological Review* 51, no. 4 (1986): 464–481.

Song Yangbiao. "The Debates on GM Rice as the Staple Food Up Again" (in Chinese). *Time Weekly*, March 11, 2010.

Sood, Jyotika. "Whose Germplasm Is It?" *Down to Earth*, October 15, 2012. Accessed August 14, 2015. http://www.downtoearth.org.in/coverage/whose-germplasm-is-it-39206.

Spector, Michael. "Seeds of Doubt: An Activist's Controversial Crusade Against Genetically Modified Crops." *New Yorker*, August 25, 2014.

State Council Information Office. "News Conference on the Acceleration of Agricultural Modernization" (in Chinese). February 3, 2015. Accessed January 30, 2016. http://www.scio.gov.cn/xwfbh/xwbfbh/wqfbh/2015/20150203/tw32541/Document/1393981/1393981.htm.

State Intellectual Property Office (SIPO). "The Argument of Foreign Companies' Control of Chinese GM Crop Patents Is Not Right" (in Chinese). *Legal Daily*, March 23, 2010, 6.

———. "News Conference to Interpret Central Document No. 1: Key Points" (in Chinese). January 28, 2016. Accessed January 30, 2016. http://www.scio.gov.cn/zxbd/wz/Document/1466802/1466802.htm.

Stone, Glenn Davis. "The Anthropology of Genetically Modified Crops." *Annual Review of Anthropology* 39 (2010): 381–390.

Stone, Richard. "Activists Go on Warpath Against Transgenic Crops—and Scientists." *Science* 331 (February 25, 2011): 1000–1001.

——. "China Plans $3.5 Billion GM Crops Initiative." *Science* 321 (September 5, 2008): 1279.

Sun Tao, Cai Tingyi, and Xu Jin. "The Abnormal Golden Rice Experiment in Hengyang" (in Chinese). *Caijing Magazine*, no. 23 (September 10, 2012).

Sun, Yutao, and Cong Cao. "Demystifying Central Government R&D Spending in China." *Science* 345 (August 29, 2014): 1006–1008.

Suttmeier, Richard P. "Engineers Rule, OK?" *New Scientist*, November 10, 2007, 71–73.

Suttmeier, Richard P., and Cong Cao. "China Faces the New Industrial Revolution: Assessing Research and Innovation Strategies for the 21st Century." *Asian Perspective* 23, no. 3 (1999): 153–200.

——. "China's Technical Community: Market Reform and the Changing Policy Cultures of Science." In *Chinese Intellectuals Between Market and State*, ed. Edward Gu and Merle Goldman, 138–157. London: RoutledgeCurzon, 2004.

Suttmeier, Richard P., Cong Cao, and Denis Fred Simon. " 'Knowledge Innovation' and the Chinese Academy of Sciences." *Science* 312 (April 7, 2006): 58–59.

——. "China's Innovation Challenge and the Remaking of the Chinese Academy of Sciences." *Innovations: Technology, Governance, Globalization* 1, no. 3 (2006): 78–97.

Tan, Monica. "24 Children Used as Guinea Pigs in Genetically Engineered 'Golden Rice' Trial." Blog post, August 31, 2012. Accessed May 31, 2014. http://www.green peace.org/eastasia/news/blog/24-children-used-as-guinea-pigs-in-geneticall /blog/41956/.

Tang, Guangwen, Yuming Hu, Shi-an Yin, Yin Wang, Gerard E. Dallal, Michael A. Grusak, and Robert M. Russell. "β-Carotene in Golden Rice Is as Good as β-Carotene in Oil at Providing Vitamin A to Children." *American Journal of Clinical Nutrition* 96, no. 3 (2012): 658–664.

Tian Jiaojin. "A Hundred or So Scholars Petition and CPPCC Members Propose: Commercialization of GM Rice Under Scrutiny" (in Chinese). *Youth Times*, March 11, 2010, A3.

Timmermans, Arco. "Accommodation, Bureaucratic Politics, and Supranational Leviathan: ART and GMO Policy-Making in the Netherlands." In *The Politics of Biotechnology in North America and Europe: Policy Networks, Institutions, and Internationalization*, ed. Éric Montpetit, Christine Rothmayr, and Frédéric Varone, 169–192. Lanham, MD: Lexington Books, 2007.

Toke, Dave. *The Politics of GM Food: A Comparative Study of the UK, USA and EU.* Abingdon, UK: Routledge, 2004.

Toke, David, and David Marsh. "Policy Networks and the GM Crops Issue: Assessing the Utility of a Dialectical Model of Policy Networks." *Public Administration* 81, no. 2 (2003): 229–251.

Toplensky, Rochelle. "European Countries Approve Weed-Killer Glyphosate for Another Five Years." *Financial Times*, November 27, 2017.

Torgersen, Helge, Jürgen Hampel, Marie-Louise von Bergmann-Winberg, Eleanor Bridgman, John Durant, Edna Einsiedel, Böjrn Fjæstad, George Gaskell, Petra Grabner, Petra Hieber, Erling Feløe, Jesper Lassen, Athena Marouda-Chatjoulis, Torben Hviid Nielsen, Timo Rusanen, George Sakellaris, Franz Seifert, Carla Smink, Tomasz Twardowski, and Merci Wambui Kamara. "Promise, Problems and Proxies: Twenty-Five Years of Debate and Regulation in Europe." In *Biotechnology: The Making of a Global Controversy*, ed. Martin W. Bauer and George Gaskell, 21–94. Cambridge: Cambridge University Press, 2002.

Trigo, Eduadro J., and Eugenio J. Cap. "Transforming Agriculture in Argentina: The Role of Genetically Modified (GM) Crops." In *Successful Agricultural Innovation in Emerging Economics: New Genetic Technologies for Global Food Production*, ed. David J. Bennett and Richard C. Jennings, 134–149. Cambridge: Cambridge University Press, 2013.

Tripp, Robert. "Transgenic Cotton and Institutional Performance." In *Biotechnology and Agricultural Development: Transgenic Cotton, Rural Institutions and Resource-Poor Farmers*, ed. Robert Tripp, 88–104. London: Routledge, 2009.

Tu, Jumin, Guoan Zhang, Karabi Datta, Caiguo Xu, Yuqing He, Qifa Zhang, Gurdev Singh Khush, and Swapan Kumar Datta. "Field Performance of Transgenic Elite Commercial Hybrid Rice Expressing *Bacillus thuringiensis* Delta-Endotoxin." *Nature Biotechnology* 18, no. 10 (2000): 1101–1104.

Tuchman, Gaye. "Telling Stories." *Journal of Communication* 26, no. 4 (December 1976): 93–97.

US-China Business Council. "USCBC Recommendations for Revisions to the 2014 Draft Catalogue Guiding Foreign Investment in Industry." December 3, 2014. Accessed December 20, 2015. https://www.uschina.org/sites/default/files/USCBC%20Recommendations%202014%20Draft%20Catalogue%20Guiding%20Foreign%20Investment%20-%20English.pdf.

U.S. Department of Agriculture (USDA), Agricultural Research Service. "Q&A: Bt Corn and Monarch Butterflies." Accessed August 10, 2014. http://www.ars.usda.gov/is/br/btcorn/index.html#bt4.

U.S. Department of Agriculture (USDA), Animal and Plant Health Inspection Service. "Petitions for Determination of Nonregulated Status." Accessed November 10, 2017. http://www.aphis.usda.gov/biotechnology/petitions_table_pending.shtml.

U.S. Environmental Protection Agency (EPA). "Regulation of Biotechnology for Use in Pest Management." Accessed November 10, 2017. https://www.epa.gov/regulation-biotechnology-under-tsca-and-fifra/epas-regulation-biotechnology-use-pest-management.

U.S. Food and Drug Administration (FDA). "Biotechnology Consultations on Food from GE Plant Varieties." Accessed November 10, 2017. http://www.accessdata.fda.gov/scripts/fdcc/?set=Biocon.

———. "Premarket Notice Concerning Bioengineered Food." *Federal Register* 66 (January 18, 2001): 4706–4738.

———. "Statement of Policy: Food Derived from New Plant Varieties." *Federal Register* 57 (May 29, 1992): 22984–23305.

U.S. Office of Science and Technology Policy (OSTP). "Coordinated Framework for Regulation of Biotechnology." *Federal Register* 50 (June 26, 1986): 23302–23393.

Varone, Frédéric, Christine Rothmayr, and Éric Montpetit. "Comparing Biotechnology Policy in Europe and North America: A Theoretical Framework." In *The Politics of Biotechnology in North America and Europe: Policy Networks, Institutions, and Internationalization*, ed. Éric Montpetit, Christine Rothmayr, and Frédéric Varone, 1–34. Lanham, MD: Lexington Books, 2007.

Varone, Frédéric, and Nathalie Shiffino. "Conflict and Consensus in Belgian Biopolicies: GMO Controversy Versus Biomedical Self-Regulation." In *The Politics of Biotechnology in North America and Europe: Policy Networks, Institutions, and Internationalization*, ed. Éric Montpetit, Christine Rothmayr, and Frédéric Varone, 193–213. Lanham, MD: Lexington Books, 2007.

Vaughan, Adam. "French Ban of Monsanto GM Maize Rejected by EU." *Guardian*, May 22, 2014.

Victor, David A. "Trade, Science, and Genetically Modified Foods." New York: Council on Foreign Relations, March 13, 2001. Accessed September 1, 2014. http://www.cfr.org/agricultural-policy/trade-science-genetically-modified-foods/p8689.

Vidal, John, and Hanna Gersmann. "Biotech Group Bids to Recruit High-Profile GM 'Ambassadors.'" *Guardian*, October 20, 2011.

Vogel, David. *The Politics of Precaution: Regulating Health, Safety, and Environmental Risks in Europe and the United States*. Princeton, NJ: Princeton University Press, 2012.

Waltz, Emily. "China's GM Rice First." *Nature Biotechnology* 28, no. 8 (2010): 8.

Wan Li. "Democratization and Scientification of Policy-Making Is an Important Subject of the Reform of the Political System" (in Chinese). *People's Daily*, August 15, 1986.

Wang Can. "Qin Zhongda, the 92-Year-Old Former Minister of Chemical Industry and a Former Member of the CCP's Central Committee, Questioned China's Biggest Overseas Acquisition" (in Chinese). Accessed August 16, 2016. http://www.thepaper.cn/newsDetail_forward_1453446.

Wang, Changyong, Zhidi Yu, and Dehui Wang. "China: Risk Assessment and Risk Management." In *Risk Assessment and Risk Management in Implementing the Cartagena Protocol: Proceedings of Asia Regional Workshop*, ed. Balakrishna Pisupati, Bhujangarao D. Dharmaji and Emilie Warner, 90–101. Homagama, Sri Lanka: Karunaratne & Sons, 2003.

Wang Dayuan. "China Is the First Country in the World to Commercialize GM Crops" (in Chinese). Blog post, May 14, 2014. Accessed June 22, 2014. http://blog.sciencenet.cn/blog-642008-794435.html.

Wang Danhong. "Shi Yigong: The Twenty-First Century Is the Century of Life Sciences" (in Chinese). *Science Times*, August 7, 2007.

Wang, Qiang. "Chinese Scientists Must Engage the Public on GM." *Nature* 519 (March 5, 2015): 7.

Wang, Shaoguang. "Changing Models of China's Policy Agenda Setting." *Modern China* 34, no. 1 (2008): 56–87.

Wang, Xueman. "Challenges and Dilemmas in Developing China's National Biosafety Framework." *Journal of World Trade* 38, no. 5 (2004): 899–913.

Wang, Yanqing, and Sam Johnston. "The Status of GM Rice R&D in China." *Nature Biotechnology* 25, no. 7 (2007): 717–718.

Waterton, Claire, and Brain Wynne. "Knowledge and Political Order in the European Environment Agency." In *States of Knowledge: The Co-production of Science and Social Order*, ed. Sheila Jasonoff, 87–108. London: Routledge, 2004.

Watson, James D. *The Double Helix: A Personal Account of the Discovery of the Structure of DNA*. New York: Atheneum, 1968.

Wei Guihong, "Research on Legislation Issue and Advice for GMOs-Safety in China" (in Chinese). In *Risk Assessment and Regulation of Genetically Modified Organism: Proceedings of the Fourth Workshop of International Biosafety Forum*, ed. Xue Dayuan, 189–196. Beijing: China Environmental Science Press, 2012.

Wen Jiabao. "Report on the Work of the Government." March 5, 2010. Accessed June 20, 2014. http://www.npc.gov.cn/englishnpc/Speeches/2010–03/19/content_1564308.htm.

Wen Jiayun. "MOA Called Off Golden Rice Experiment in China in 2008" (in Chinese). Blog post, August 31, 2012. Accessed May 9, 2014. http://blog.sina.com.cn/s/blog_81b7cbe801014kt5.html.

Wen Jin. "Zhangye in Gansu Issued Document Banning GM Crop Growing" (in Chinese). *Jinghua Times*, November 1, 2013, 3.

Winerip, Michael. "You Call It a Tomato?" *New York Times*, June 24, 2013.

Wong, Joseph. *Betting on Biotech: Innovation and the Limits of Asia's Developmental State*. Ithaca, NY: Cornell University Press, 2011.

Wu, Kong-Ming, Yan-Hui Lu, Hong-Qiang Feng, Yu-Ying Jiang, and Jian-Zhou Zhao. "Suppression of Cotton Bollworm in Multiple Crops in China in Areas with Bt Toxin–Containing Cotton." *Science* 321 (September 19, 2008): 1676–1678.

Wu Shihong. "Deng Xiaoping's Thinking on Modernization of Agriculture in China" (in Chinese). *Literature of the Chinese Communist Party*, no. 6 (November 2006): 61–66.

Xi Jinping. "China Should Occupy the Commanding Heights of Transgenic Technology" (in Chinese). September 28, 2014. Accessed September 28, 2014. http:// politics.caijing.com.cn/20140928/3711853.shtml.

Xia Shang. "Science Communication Behaviors of Greenpeace China on Genetically Modified Organism Issues" (in Chinese). Master's thesis, Peking University, Beijing, 2007.

Xinhua News Agency. "Heilongjiang Will Forbid Planting GM Food Crops in Next Years" (in Chinese). December 26, 2016. Accessed June 10, 2017. http://news .xinhuanet.com/politics/2016–12/16/c_1120134366.htm.

Xu Haigen, Li Wei, Lu Yan, and Zheng Yanping. "The Construction of Biosafety Clearing House of China and Its Proposal." In *Environmental Impacts and Safety Regulation of Genetically Modified Organisms: Proceedings of the International Biosafety Forum*, ed. Xue Dayuan, 96–97. Beijing: China Environmental Science Press, 2006.

Xue, Dayuan. *A Summary of Research on the Environmental Impacts of Bt Cotton in China*. Hong Kong: Greenpeace, 2002.

Yan Dingfei, and Huang Boxin. "Thousand Persons Try GM Rice: Is It a Serious Activity of Science Popularization" (in Chinese). *Southern Weekend*, July 26, 2013.

Yang, Guobin, and Craig Calhoun. "Media, Civil Society, and the Rise of a Green Public Sphere in China." *China Information* 21, no. 2 (2007): 211–236.

Yang, Jinjie, Kaibin Xu, and Lulu Rodriguez. "The Rejection of Science Frames in the News Coverage of the Golden Rice Experiment in Hunan, China." *Health, Risk & Society* 16, no. 4 (2014): 339–354.

Yap, Chuin-Wei. "It's China vs. China in Genetically Modified Food Fight." *China Real Time Report* (blog). *Wall Street Journal*, September 3, 2013. Accessed August 18, 2014. http://blogs.wsj.com/chinarealtime/2013/09/03/its-china-vs-china-in-gmo -food-fight/.

Yin Shuai-Jun. "Social Risk Management and Control for GMOs" (in Chinese). In *Risk Assessment and Risk Management of the Genetically Modified Organisms: Proceedings of the Fifth Workshop of the International Biosafety Forum*, ed. Xue Dayuan, 107–119. Beijing: China Environmental Press, 2014.

Young, Alasdair R. "Political Transfer and 'Trading Up'? Transatlantic Trade in Genetically Modified Food and U.S. Politics." *World Politics* 55, no. 4 (2003): 457–484.

Yu Dawei. "GM Rice: The Eve of Its Commercialization" (in Chinese). *New Century Weekly*, no. 8 (February 22, 2010). Accessed February 17, 2010. http://magazine .caixin.com/2010-02-17/100117846.html.

Yu Jingyin. "It's Safe to Eat Imported GM Soybean Oils" (in Chinese). *People's Daily* (Overseas Edition), June 21, 2013.

Yu, Jun. "Biotechnology Research in China: A Personal Perspective." In *Innovation with Chinese Characteristics: High-Tech Research in China*, ed. Linda Jakobson, 134–165. New York: Palgrave McMillan, 2007.

Yu Wenxuan, and Chen Yi. "China's Agro-GMO Biosafety Legislation and Its Improvement." In *Risk Assessment and Regulation of Genetically Modified Organisms: Proceeding of the Fourth Workshop of the International Biosafety Forum*, ed. Xue Dayuan, 177–188. Beijing: China Environmental Science Press, 2012.

Yuan Yue. *Analysis of News of Genetically Modified Crops and Foods in Four Chinese Newspapers* (in Chinese). Master's thesis, Graduate School of the Chinese Academy of Sciences, Beijing, 2008.

Zaller, John R. *The Nature and Origins of Mass Opinion*. Cambridge: Cambridge University Press, 1992.

Zhang Luyu. "Editor's Note: Closing to the Truth Makes People So Worry" (in Chinese). *International Herald Leader*, September 21, 2010.

Zhang Qifa. "China." In *Agricultural Biotechnology: Country Case—A Decade of Development*, ed. Gabrielle J. Persley and L. Reginald MacIntyre, 41–50. Wallingford, UK: CABI, 2001.

Zhang Qifa et al. "Suggestions on the Development Strategy of Research and Commercialization of GM Crops in China" (in Chinese). *Bulletin of the Chinese Academy of Sciences* 19, no. 5 (2004): 330–332.

Zhang Siguang, Shen Xiaobai, and Duan Yibing. "Public-Private Partnerships in Commercialization of Biotechnology: The Case of Transgenic Rice in China" (in Chinese). *Scientific Research Management* 27 (2006): 71–76.

Zhang, Tao, and Shudong Zhou. "The Economic and Social Impact of GMOs in China." *China Perspectives*, no. 47 (June 2003): 50–57.

Zhang Wei. "'Covert' GMOs" (in Chinese). *China Youth Daily*, December 23, 2009.

Zhang Weibin. "Who Defeat Scientists on GMOs?" (in Chinese). *Hubei Daily*, December 2, 2013.

Zhang Zhongfa. "Commercialization and Safety of Transgenic Biotechnology" (in Chinese). In *China Development Studies 2002: Selected Research Reports of Development Research Center of the State Council*, ed. Ma Hong and Wang Mengkui, 514–522. Beijing: China Development Press, 2002.

Zhao, Juan. "News Coverage of GM Crops in China: Problems and Solutions" (in Chinese). *Forum of Chinese Association of Science and Technology*, no. 2 (2012): 191–192.

Zhao Zuo. "Monsanto Phantom: Crisis on Chinese's Dining Table" (in Chinese). *Time Weekly*, July 8, 2010.

Zhong, Funing, Mary A. Marchang, Yulian Ding, and Kaiyu Lu. "GM Food: A Nanjing Case Study of Chinese Consumers' Awareness and Potential Attitude." *AgBioForum* 5, no. 4 (2003): 136–144.

Zhou, Ping, and Wolfgang Glanzel, "In-depth Analysis on China's International Collaboration in Science." *Scientometrics* 82, no. 3 (2010): 597–612.

Zhou, Yuqiong, and Patricia Moy. "Parsing Framing Processes: The Interplay Between Online Public Opinion and Media Coverage." *Journal of Communication* 57, no. 1 (2007): 79–98.

Zhou Yijun, Xue Kun, Liu Qing, Hou Liting, and Xue Dayuan. "Investigation on the Intellectual Property Rights About Transgenic Bt Rice" (in Chinese). In *Risk Assessment and Safety Regulation of Genetically Modified Organisms: Proceedings of the Third Workshop of the International Biosafety Forum*, ed. Xue Dayuan, 133–169. Beijing: China Environmental Science Press, 2009.

Zhou Yingfeng. "Xi Jinping Participated in Popularization of Science Activities at China Agricultural University" (in Chinese). *Xinhua News Agency*, September 15, 2012. Accessed September 28, 2014. http://www.gov.cn/ldhd/2012–09/15/content_2225485.htm.

Zhu Rongji. "Develop Ourselves by Fully Utilizing the Opportunity of WTO Accession" (in Chinese). February 24, 2002. In *Zhu Rongji on the Record*, comp. Editorial Board, 4: 312–323. Beijing: People's Press, 2011.

Ziman, John. *Real Science: What It Is and What It Means.* Cambridge: Cambridge University Press, 2000.

——. "Research as a Career." In *The Research System in Transition*, ed. Susan E. Cozzens, Peter Healey, Arie Rip, and John Ziman, 345–359. Dordrecht: Kluwer, 1990.

Index

Page numbers in italics indicate figures or tables.

Argentina, 25, 197
Aventis, 9–10

Bacillus thuringiensis (Bt), 5
BASF, 25, 27, 101
Bayer, 10, 29; acquisition of
    Monsanto, 163
Beijing Genomics Institute, 58
Ben and Jerry's, 22
benefits, 136, *137*, 175
Berg, Paul, 2–3
Biocentury, 71, 90
biodiversity, 82; Greenpeace and, 86;
    safety concerns on, 126; scientists
    worried about, 190; Xue as expert
    on, 63, 86, 91, 113
biofuels, 1
biomedical, 8
biosafety certificates, xviii, 56, 57, 191;
    approval stages for, 159; Bt rice with,
    103–104, 108; GM rice waiting for,
    159, *160*; as not granted, 69; scholars
    for withdrawing, 109–110
biosafety regulatory regime, xvii–xviii,
    46, 63, 65, 76; Bt rice certification in,
    92–95; Chinese indigenous genes not
    approved in, 71–72; debates centered
    on, 184; evolution of, 77–82, *78–79*;
    government with rigorous, 101–102,
    195; Greenpeace concerns on, 126;
    health commission for, 85–86; India
    MOST and DBT approving, 30–31;
    MOA inspecting for, 80, 85, 98, 181;
    1997 committee in charge of, 80;
    operation of, 85–91, *88*; provinces
    role in, 86; public and media
    resulting in, 181; rationales for, 82–84;
    regulations in, 95–99; trade barrier
    perception from, 182–183; 2001
    State Council regulation of, 80–81;
    untested seeds without laws in, 103

biotechnology, 1, 3–4, 8, 13, 16; agro-
    food as green and undesirable, 130;
    economic strategy with, 39; India
    states monitoring, 31; India with Bt
    cotton, pharmaceuticals, services in,
    32; investments in, 58–59; medical
    as "red" and desirable, 130; 1989 and
    1997 with science and research for,
    54–55; policy agendas developing,
    51–57; programs for, *52–53*; U.S. gap
    with EU, 25. *See also* agricultural
    biotechnology
Boyer, Herbert W., 2–3
Bt (*Bacillus thuringiensis*), xv
Bt brinjal (eggplant), 32, 33–35
Bt cotton, 70–71; bollworm and,
    73, 113; as domestic crop, 64–65,
    67, 69–70, 196; domestic and
    agriculture newspapers for, 135–136;
    illegal planting of, 96; India and,
    31–32, 35–36; indigenous varieties
    of, 196; IPR landscape for, 174–175;
    patents and success on, 162; pesticide
    use with, 72, 113; prices reduced by,
    73–74
Bt European black poplar trees, 66
Bt papaya, 67
Bt rice, 67, 69, 102; bashful policy
    released quietly on, 95, 149–150;
    biosafety certificates on, 103–104,
    108; biosafety regulatory for
    certifying, 92–95; certification
    controversy on, 92, 155;
    development of, 93–94, *165*;
    Greenpeace and, 94–95, 97–98,
    114–115; Greenpeace with IPR
    questions on, 156–157; indigenous
    genes for, 157–158; IPR ownership
    for, 111; media coverage of, 134–135,
    147; newspapers limiting coverage
    of, 147–148; patent portfolio analysis

of, 166–173, *168*, *169*, *171*; public concerns on, 92, 94–95, 104, 134, 155; scientists and process for, 99, 108, 157. *See also* GM rice

Bt Shanyou 63 (rice), 163, 166, 174; benefit-sharing risks to, 172–173; Bt rice with certificate as, xviii, 68, 92, 147, 164; composition of, 164, *165*; Cry genes and, 159; economic benefits not ensured by, 175; five of six patents implied in, 167, *168*; foreign patent trap for, 167, 170; indigenous techniques for, 157–158; as unauthorized ingredient, 98–99

bureaucratic policy culture, 191–192

CAE. *See* Chinese Academy of Engineering
Calgene, 6, 67
carcinogen, 11
CAS. *See* Chinese Academy of Sciences
Cas9, 13
CASS. *See* Chinese Academy of Social Sciences
Cartagena Protocol on Biosafety, 82, 86, 182
CBD. *See* Convention on Biological Diversity, UN
CCAP. *See* Center for Chinese Agricultural Policy
CCP. *See* Chinese Communist Party
CCPCC. *See* Chinese Communist Party's Central Committee
CCTV. *See* China Central Television
Center for Chinese Agricultural Policy, 61, 65
central document (*zhongyang wenjian*), 44
Central Document No. 1, 176
Central Television (CCTV), China, xv

certification process, 109; Bt rice controversy on, 155; Bt Shanyou 63 approval in, xviii, 68, 92, 147, 164
ChemChina. *See* China National Chemical Corp
Chen Xiwen, 48
Chen Zhangliang, 60–61, 66, 121
Chien-Shiung Wu, 54
China. *See specific subjects*
China Agricultural University, 50, 60, 91, 94, 111, 199
China Association for Science and Technology, 61, 111, 124
China National Chemical Corp (ChemChina), 5, 29, 199
China National Knowledge Infrastructure (CNKI), 131–132, 167
China National Rice Institute, 121
Chinese Academy of Agricultural Sciences (CAAS), 56, 62, 66; Institute of Biotechnology of, 67, 69, 174–175
Chinese Academy of Engineering (CAE), 43, 56
Chinese Academy of Sciences (CAS), 43, 56, 59, 62
Chinese Academy of Social Sciences (CASS), Institute of Rural Development under, 63
Chinese Centers for Disease Control and Prevention (CDC), 85, 94, 117, 125
Chinese Communist Party (CCP), 41–42
Chinese Communist Party's Central Committee (CCPCC), 41–42; Politburo Standing Committee of, 41;
Chinese People's Political Consultative Conference (CPPCC), xv, 40, 110–111, 177

Chipotle Mexican Grill, 22
civic policy culture, 192–193
CJD. *See* Creutzfeldt-Jacob Disease
Climbing Program (*pandeng jihua*), 54
closed-door policymaking, 42
clustered regularly interspaced short
palindromic repeats (CRISPR), 13
Cohen, Stanley N., 2–3
commercialization, xviii, 65, 150–151;
becoming remote, 195; food safety
complicated by, 193; GMO research
theme with, *137*, 137–138; IPR
concerns on, 175–176; licenses issued
for, 7, 66–67, *68*; MOA approving,
31, 77
Communication University of
China, xv
conflict of interests, 90–91, 96;
committee members with GMO,
109–110; corporate funding as, 187;
science entrepreneurs with, 54, 74,
178–180, 187, 189; scientist personal
interests as, 13, 105, 184; weeklies
detailing, 149–150
conspiracy theory: American plot fears
as, xvi–xvii, 105, 111–112, 115, 118–
123, 190; Chinese biotechnologists
and, 106, 123; Engdahl and, 118–120;
Ford Foundation in, xvii, 119–121;
GM crops versus food security in,
122–123; Monsanto and, 120–121,
184–185, 190; political tolerance of,
127; rice tasting in, 124–125
Consultative Group on International
Agricultural Research (CGIAR),
61, 121
conventional breeding, 13
Convention on Biological Diversity
(CBD), UN, 63, 82, 172–173, 182
Cornell University, xvii, 9, 54, 59–62,
122, 167

corporate funding, 54, 74, 178–180,
187, 189
CPPCC. *See* Chinese People's Political
Consultative Conference
Creutzfeldt-Jacob Disease (CJD), 22–23
Crick, Francis, 2
CRISPR. *See* clustered regularly
interspaced short palindromic
repeats
CRISPR/Cas9, 13, 198–199; U.S. and
EU regulations converging with, 37
Cry genes: Cry1Ab, 159; Cry1Ac, 33,
34, 164, *165*; Cry9c, 9, 159
Cui Yongyuan, xv, 90, 177
Cultural Revolution, 51, 57–58

DBT. *See* Department of
Biotechnology, India
Deng Xiaoping, 51, 54
deoxyribonucleic acid (DNA), 2, 4
Department of Biotechnology (DBT),
India, 30–31
diagnostic toolkits, 1
DNA. *See* deoxyribonucleic acid
DNA insertion, 4
domestic stakeholders, xviii
DuPont, 123; DuPont Pioneer, 62,
121, 152

economic policy culture, 192
economic strategy, 39
education, 59–60
EFSA. *See* European Food Safety
Authority
emerging technology/technologies, xix,
140, 153, 186
Endangered Species Act (ESA),
U.S., 19
Engdahl, F. William, 118–120
entrepreneurs, 54, 74, 178–180, 189
Environment Support Group, 35

environmental benefits, 12, 73;
biosafety regulatory regime and, 86;
transgenic with uncertain, 179

Environmental Protection Act of 1986
(EPA 1986), India, 30

Environmental Protection Agency
(EPA), U.S., 9, 11; glyphosate and
herbicides regulated by, 18; GMO
impact complaint by, 20; new PIP
and GMMP categories from, 19;
pesticide regulation guidance for, 17

environmental risks, 8, 17, 19, 20;
EU concern on, 22; unintended
consequences as, 13

EPA 1986. *See* Environmental
Protection Act of 1986, India

equivalency, 16, 28, 50; labeling
contrasted with, 20; U.S. approval of
button mushroom as, 37; U.S. FDA
ruling for, 8, 17–18

ESA. *See* Endangered Species Act, U.S.

ethics, 12, 116; civic policy structure
and, 192; and policy considerations
in journalism, 151; Tufts criticized on
Golden Rice, 117

EU. *See* European Union

European Food Safety Authority
(EFSA), 24, 28

European Union (EU), xvii, 100; anti-
GMO activists in, 27; anti-GMO
activists similar with, 105; biosafety
and exports to, 84; CRISPR/Cas9
for U.S. regulation convergence
with, 37; food safety scandals
influencing, 196; GM crops and
U.S., 36; GM policy in, 22–30; GM
preventive attitude in, 15, 22, 37,
82, 196

exports, 84, 183; EU and crop
rejections in, 15; illegal planting
influencing, 99

Fan Yunliu, 62, 93, 121

Fang Lifeng, 114

Fang Yi, 54

Fang Zhouzi, xv, 108, 112, 124–125

farmers, 12, 162; Bt brinjal concern
and indigenous medicine by, 34; Bt
cotton costs for, 72–73; Bt rice and
frustration of, 99; protectionism and
Chinese soybean, 83–84

Farmer's Daily, 138

FDA. *See* Food and Drug
Administration, U.S.

fenzi yuzhong (molecular breeding),
47–48

first central document issued in year
(*zhongyang yihao wenjian*), 44

Flavr Savr, 5–6

Food and Drug Administration (FDA),
U.S., 8, 17–19, 48

food consumer rights, xviii

Food Law, 198

food safety, xviii, 184; EU influenced
by, 196; GM commercialization
complicating, 193; Greenpeace
addressing, 113; Greenpeace
concerns on, 126; IPR debate
on, 186

Food Safety and Standards Authority
(FSSA), India, 31

food-safety scandals, 189, 196; Chinese,
109; CJD as, 22–23; EU science,
business, regulators distrusted in,
26–27; Golden Rice as, 115–118

food security, xviii, 39, 176; conspiracy
theory on GM crops and, 122–123;
debates centered on, 184; GM crops
for, 185; health and environment
versus, 102

Ford Foundation, 61, 120–121

foreign patent trap, 156–157, 167, 170

Fresco, Louise O., 28

Leading Group for Science, Technology, and Education, 41–42, 55

LibertyLink rice, 10, 36–37, 186

Li Jiayang, 62, 121–122

Li Keqiang, 50

litigation risks, 19

live modified organism (LMO), 31, 84, 167–168, 173

Lu Baorong, 118

Lynas, Mark, 29–30, 37

media coverage, xix; of Bt rice, 135, 147; policy shaped by, 129–130; precaution shaped by, 180; public influenced by, 153; transgene trends and, 132, *133*, *134*; 2002 and GM terms in, 132–133; 2009 with Bt rice and phytase maize stories in, 134

medical devices, 1

Medium- and Long-Term Plan for the Development of Science and Technology (2006–2020), 43–44, 55, 56, 64, 65, 141

Mega-Engineering Program (MEP), China, 44, *53*, 57, 74, 108, 159, *191*

Mega-Science Program (MSP), China, 55

Mendel, Gregor, 2

MEP. *See* Mega-Engineering Program, China

Mertz, Janet E., 2

Ministry of Agriculture (MOA), China: biosafety regime inspections by, 80, 85, 98, 181; Bt rice and position by, 102; commercial release approval by, 31, 77; policies as challenged, 196

Ministry of Commerce (MOFCOM), 79, 83, 85

Ministry of Environmental Protection (MOEP), China, 182

Ministry of Science and Technology (MOST), China, 41, 51, *53*, 59, 77, 85, 102, 131

Ministry of Science and Technology (MOST), India, 30–31

MLP. *See* Medium- and Long-Term Plan for the Development of Science and Technology (2006–2020)

MOA. *See* Ministry of Agriculture, China

MOEP. *See* Ministry of Environmental Protection, China

molecular breeding (*fenzi yuzhong*), 47–48

monarch caterpillars, 9, 36

Monsanto, 114, 123; anti-GMO activists charges on, 120–121; Bayer taking over, 29; Bt cotton varieties from, 72; Chinese Bt cotton and, 71; Chen Zhangliang and, 60–61, 121; conspiracy theory and, 184–185, 190; EFSA approval of soybean from, 28; gene-gun transformation by, 167, *168*; improving Bt cotton and, 70; joint ventures with, 178; labeling and millions spent by, 21; new seeds not introduced by, 196; patents not pursued by, 186; Shiva and seed cost by, 35–36; SIPO submissions from, 163; Zhang Qifa and, 61, 93, 121

MOST. *See* Ministry of Science and Technology, India; Ministry of Science and Technology, China

Nagoya Protocol on Access to Genetic Resources and Benefit-Sharing, 172–173

Nanjing Institute of Environmental Sciences, 63, 86, 112–113

National Academy of Sciences (NAS), U.S.

Administration, China, 85, 116, 124, 157
Office of Science and Technology Policy (OSTP), U.S., 15, 81, 100
organic farming, xvi, 179, 196, 198
Organization for Economic Co-operation and Development (OECD), 14–15
Origin Agritech, 67, *68*, 71
OSTP. *See* Office of Science and Technology Policy, U.S.
Outlook Weekly, 150

patent portfolio analysis: of Bt rice, 156; Chinese Bt rice in, 166–173, *168*, *169*, *171*; Greenpeace confirmation in, 170
patent protection: foreign patent trap and, 156–157; upstream and licenses for, 167, *169*
patents, xix; applications as increasing, 161, *161*; Bt rice questions on, 156; Bt Shanyou 63 and implied, 167, *168*; limitations to, 170, *171*; Monsanto not pursuing China, 186; technology and materials for, 164–166, *165*; universities and research receiving, 161–162, *163*; weeklies detailing influence of, 149–150
Peng Guangqian, 122, 185
Peng Yufa, 63, 91, 95, 144
Peng Zhongming, 93, 164
Peking University, 67, *68*
People's Daily, 138
permissive attitude, xvii
pesticides: Bt cotton and, 72, 113; Bt Shanyou 63 and, 164, *165*; environmental benefits and reduced, 73; U.S. EPA regulation guidance for, 17; GMMP as, 17, 113; insect tolerance for, 4–5; resistant crops and, 184

pharmaceuticals: GM and research in, 25–26; "red" medical biotechnology as, 130
plant-incorporated protectants (PIPs), 17, 19
plant-variety right, 173, 175
policy agendas, xviii–xx; anti-GMO influencing, 177; for biotechnology development, 51–57; debates centered on, 184; democratization of, 42; elites proposing petitions on, 44; EU core of, 23–25; EU with preventive, 22; global debates influencing, 103, 194; as global transgenic, 38; Golden Rice and hesitation on, 154; government as ambivalent on, 151, 193; government with precaution, xvii, 46, 102; on hydropower, 193–194; India as precautious and preventative on, xvii, 15; media coverage shaping, 129–130; more stakeholders in, 188–189; newspaper tone set by, 143; OECD countries diverging on, 14–15; pro- and anti-GMO criticism of U.S., 19–20; science impacts and, 186–187; scientists informed by, 197–198; stakeholders involvement in, 13; 2012 shift to precaution, 149; U.S. business shaping, 178; U.S. importation, interstate movement, field-testing, 17; U.S. without stringent, 16
policymaking, agenda-setting, 42-43
political actions, xvi, 187; Bt brinjal and consensus of, 34; conspiracy theory tolerance as, 127; Greenpeace mobilizing for, 114
pollen drift, 10, 36–37
pollen-tube pathway transformation, 4, 174–175

regulations, 3, 195; biosafety and compromised, 95–96, 102; Chinese and international, 99–101; framework for, *78–79*; GM tobacco before development of, 77, *78–79*; Golden Rice study increasing, 116–117; right to know theme with, *137*, 138

regulatory processes, 56

regulatory sequence of DNA, 4

research, xviii, 64–66; funding decreased on, 191, *191*; GM rice, 159–163, *160, 161, 163*; government supporting, 195; IPR analysis and, 175; open-door policy making and, 42; questions on objectivity of, 13, 105, 184

retailers, 27

retractions, 8–10

returnee, 59, 60, 71

ribonucleic acid (RNA), 13

right to know (*zhiqingquan*), 48–49, 83, 106–108, 111, 190; consumers with, 179–180; Greenpeace addressing, 113; safety and regulations theme with, *137*, 138

risk assessment, 8, 65, 147, 198; biosafety regime addressing, 101–102, 195; Bt cotton and food questions in, 70; Bt Shanyou 63 and, 172–173; civic policy structure and, 192; EU credibility on, 23; gene flow as nontarget in, 11, 126, 184; IPR need of, 175; NBCAGMO responsible for, 87; Nestlé amplifying, 145–146; 1997 committee in charge of, 80; rDNA evaluations of, 3, 14; reconciling perceptions of, 188; SEPA and, 86; U.S. and UK newspapers on, 136, *137*; Xue Dayuan active in, 63, 86, 91, 113

RNA. *See* ribonucleic acid

Robin, Marie-Monique, 119

Rockefeller Foundation, xvii, 61, 63, 119–121

Roundup herbicide, 4–5

Roundup Ready crops, 4–5

Rowett Institute, 8–9

Royal Society of London, 9

Rozelle, Scott, 62

safety, 18, 50; assessment on, 48–49; biodiversity concerns in, 126; Bt brinjal testing for, 33–34; Bt cotton bypassing tests for, 70; Bt rice and assessment of, 94; EU percentage concerns over, 26; public concerns on, 109; right to know theme with, *137*, 138. *See also* biosafety regulatory regime; food safety

Science, 148

science and technology (S&T), xix–xx, 43, 134; CCP with final say on, 41–42; China Cultural Revolution influence on, 57–58; China NPC for laws on, 40; GMO policy and, 191–194; GM stories and sections for, 135–136, *136*; leadership by 2050 in, 55; 1975 and 1986 plans with genetic engineering as, 51, *52–53*

Science and Technology Daily, 138, 148

science-based approach, 18

science journals, 8–10, 140

ScienceNet, 111

scientific community, 12, 23, 27, 43; Central Documents No. 1 assisting, 48; China and international, 75; China and respect of, 49; concerns from EU, 28; corporate funding as problem for, 54, 74, 178–180, 187, 189; dialogue with citizens by, 126; policies informing, 197–198; science explained by, 124 scientists, xvii–xix, 8, 56–57, 64; on biodiversity, 190; Bt brinjal concern and indigenous medicine by, 34; Bt rice and, 99, 108, 157; competitiveness and Chinese-American, 51, 54; conspiracy responses by, 124; conspiracy theories demonizing, 106, 121, 123; as entrepreneurs, 54, 74, 178–180, 189; GM benefitting from China, 74; Golden Rice experiment by, 115–116; governments interconnected with, 189; international links for Chinese, 60; interview sources as, *139*, 140; journalists and negative views by, 152; newspapers with opinions from, 143, *144*; objectivity questioned of, 13, 105, 184; policy agendas informing, 197–198; policy set by, 178; precaution shift blame and, 103; public concerns feud and, 105; reform and open-door era benefiting, 127; transgenic and nontransgenic, 179

second-generation GM crops, 5, 28, 65

SEPA. *See* State Environmental Protection Administration, China

Séralini, Gilles-Eric, 10–11, 28

Shiva, Vandana, 35–37

SIPO. *See* State Intellectual Property Office, China

Smith, Jeffrey M., 118

social media, xvii, 129–130, 180; biosafety regulatory regime from, 181; conspiracy theories and, 123

society protection, 189

South China Agricultural University, 67, *68*

Southern Weekend, 149–150

soybeans, 83–84; China largest importer of, 196–197

SSTC. *See* State Science and Technology Commission, China

S&T. *See* science and technology

Stanford University, 2–3, 36, 62

StarLink, 9, 20, 36

State Basic Research and Development Program, *53*, 54–55, 216*n*42

State Council, China, 40–41, 81–82, 85, 87, 181; ministries under, 40; State Leading Group for Science, Technology, and Education within, 41

State Environmental Protection Administration (SEPA), China, 86, 112–113, 181–182

State High-Tech Research and Development Program, 43, 51, *52*, 54, 58–59, 215*n*19

State Intellectual Property Office (SIPO), China, 158, 163

State Science and Technology Commission (SSTC), China, 99–100, 181

strategic research, 42

substantial equivalence. *See* equivalency

Sun Zhengcai, 122

superweeds, 83, 184

Supreme Court, India, 32–33, 35

Syngenta, 25, 123, 163, 199; ChemChina acquiring, 5, 29

Sze Pang Cheung, 106

Tang, Guangwen, 115, 117, 118

tasting sessions, 124–125

terminator, 4

Third World Network, 86, 156–158, 163, 172–176

Tong Pingya, 120–121

traceability, 24

trade barriers, 182–183; U.S. and Argentina suing EU over, 25

yields, 111

yuanshi (elite members), 43–44, 129

Zhang Qifa, 55–56, 61, 163; attack on, 111–112; Greenpeace holding liable, 97–98

Zhejiang Academy of Medical Sciences, 115–116

zhiqingquan (right to know), 48–49, 106, 138, 179

Zhongguancun Science Park, 50, 56

zhongyang yihao wenjian (first central document issued in year), 44

Zhou Guangyu, 4

Zhou Guangzhao, 4, 174

Zhou Li, 122

Zhou Peiyuan, 40

Zhu Rongji, 56, 64, 70, 74, 84

Zhu Zhen, 62, 63, 121

zhuan-jiyin (transgenes), xv, 118

GPSR Authorized Representative: Easy Access System Europe, Mustamäe tee
50, 10621 Tallinn, Estonia, gpsr.requests@easproject.com

www.ingramcontent.com/pod-product-compliance
Lightning Source LLC
Chambersburg PA
CBHW022139020426
42334CB00015B/972